This series aims to report new developments in physical research and teaching — quickly, informally, and at a high level. The type of material considered for publication includes:

1. Preliminary drafts of original papers and monographs

2. Lectures on a new field, or presenting a new angle on a classical field

3. collections of seminar papers

4. Reports of meetings

Texts which are out of print but still in demand may also be considered if they fall within these categories.

The timeliness of a manuscript is more important than its form, which may be unfinished or tentative. Thus, in some instances, proofs may be merely outlined and results presented which have been or will later be published elsewhere.

Publication of *Lecture Notes* is intended as a service to the international physical community, in that a commercial publisher, Springer-Verlag, can offer a wider distribution to documents which would otherwise have a restricted readership. Once published and copyrighted, they can be documented in the scientific libraries.

Manuscripts

Manuscripts are reproduced by a photographic process; they must therefore be typed with extreme care. Symbols not on the typewriter should be inserted by hand in indelible black ink. Corrections to the typescript should be made by sticking the amended text over the old one, or by obliterating errors with white correcting fluid. The figures (in the original size) ready for reproduction should be inserted into the text. Should the text, or any part of it, have to be retyped, the author will be reimbursed upon publication of the volume. Authors receive 50 free copies.

The typescript is reduced slightly in size during reproduction, therefore a large size of type should be used; best results will not be obtained unless the text on any one page is kept within the overall limit of 18 x 26.5 cm (7 x 10½ inches). The publishers will be pleased to supply on request special stationery with the typing area outlined.

Manuscripts in English, German or French should be sent to Springer-Verlag, 6900 Heidelberg, Postfach 1780.

Die „*Lecture Notes*" sollen rasch und informell, aber auf hohem Niveau, über neue Entwicklungen in der Physik berichten. Zur Veröffentlichung kommen:

1. Vorläufige Fassungen von Originalarbeiten und Monographien.

2. Spezielle Vorlesungen über ein neues Gebiet oder ein klassisches Gebiet in neuer Betrachtungsweise.

3. Seminarausarbeitungen.

4. Vorträge von Tagungen.

Ferner kommen auch ältere vergriffene spezielle Vorlesungen, Seminare und Berichte in Frage, wenn nach ihnen eine anhaltende Nachfrage besteht.

Die Beiträge dürfen im Interesse einer größeren Aktualität durchaus den Charakter des Unfertigen und Vorläufigen haben. Sie brauchen Beweise unter Umständen nur zu skizzieren und dürfen auch Ergebnisse enthalten, die in ähnlicher Form schon erschienen sind oder später erscheinen sollen.

Die Herausgabe der „*Lecture Notes*" Serie durch den Springer-Verlag stellt eine Dienstleistung an die physikalischen Institute dar, indem der Springer-Verlag für ausreichende Lagerhaltung sorgt und einen großen internationalen Kreis von Interessenten erfassen kann. Durch Anzeigen in Fachzeitschriften, Aufnahme in Kataloge und durch Anmeldung zum Copyright sowie durch die Versendung von Besprechungsexemplaren wird eine lückenlose Dokumentation in den wissenschaftlichen Bibliotheken ermöglicht.

Lecture Notes
in Physics

Edited by J. Ehlers, München, K. Hepp, Zürich and
H. A. Weidenmüller, Heidelberg
Managing Editor: W. Beiglböck, Heidelberg

13

Michael Ryan

University of North Carolina, Chapel Hill
and
University of Maryland, College Park

Hamiltonian Cosmology

Springer-Verlag
Berlin Heidelberg GmbH 1972

ISBN 978-3-540-05741-3   ISBN 978-3-540-37146-5 (eBook)
DOI 10.1007/978-3-540-37146-5

© by Springer-Verlag Berlin Heidelberg 1972.

Originally published by Springer-Verlag Berlin Heidelberg New York in 1972.

Library of Congress Catalog Card Number 77-189456.

Offsetdruck: Julius Beltz, Hemsbach/Bergstr.

PREFACE

Since the initial application of Hamiltonian techniques to cosmology a few years ago, the field has grown so rapidly that a review of the work that has been done, along with an explanation of work in progress, and an indication of possible future direction, has become desirable. It is hoped that this report will serve this purpose. The rapid publication of the LECTURE NOTES IN PHYSICS should insure that the information will reach the hands of interested people before the rapid development of the field makes it obselete.

In a work of this sort it seems almost superfluous to thank those who have made it possible; one's indebtedness stands out on every page. I would, however, like to take special note of the help, through discussions, of Prof. C. Misner, Prof. B. DeWitt, Dr. H. Zapolsky, Dr. K. Jacobs, Dr. Y. Nutku, Messrs. D. Chitre, L. Fishbone, and V. Moncrief. I wish to thank all of these people, as well as Mr. L. Hughston, Miss B. Kobre, and Dr. K. Kuchař for allowing me to see pre-publication drafts of their work.

This report was begun at the University of Maryland and approximately half of it was finished there. The second half was completed after a move to the University of North Carolina.

The format of LECTURE NOTES IN PHYSICS makes the typing and illustration of manuscripts vital. Because of this I want to express my sincere gratitude to Mrs. B. Alexander and Mrs. J. Alexander of the University of Maryland, and Mrs. J. McCloud of the University of North Carolina for their patience and the beautiful typing job they have done. I would also like to thank Mrs. J. Hudson for a set of excellent figures.

I am grateful for financial support from NASA grant
NGR 21-002-010 at the University of Maryland and an NSF grant
to Prof. B.S. DeWitt at the University of North Carolina.

MATHEMATICAL FOREWORD

While conventional symbols are used throughout for most quantities (eg. $R^{\mu}{}_{\nu\alpha\beta}$ for the Riemann tensor), most of the mathematical operations are carried out in non-coordinate frames and much of the notation is that of modern differential geometry. We give below a list of references for those unfamiliar with techniques and notation. There is a possibility of confusion between some of the notation of the calculus and that of modern differential geometry (for instance, d is used in its usual sense (dx/dt) and also to represent the operator of exterior differentiation), but the meaning of symbols should be clear from context if one is sufficiently familiar with differential geometry.

Two special points of notation need to be noted. We shall distinguish differential one-forms (but not n-forms for $n>1$) by writing them with a tilde (i.e. $\tilde{\sigma}$), and vectors (in both the modern differential geometry sense and the usual sense) by superior arrows (i.e. $\vec{v}$). Second, we shall use the notation dA/dt for a matrix A to mean

$$\frac{dA}{dt} = \frac{1}{2} [A^{-1} \frac{dA}{dt} + \frac{dA}{dt} A^{-1}] \quad (dA = \frac{1}{2} \{A^{-1}(dA) + (dA)A^{-1}\}).$$

References for Modern Differential Geometry:

See C. Misner in *Astrophysics and General Relativity (Vol. I)*, edited by M. Chretien, S. Deser and J. Goldstein (Gordon and Breach, New York, 1969) for the immediate antecedents to the ideas and notation used in the present work.

VI

The classic work is: E. Cartan, *Leçons sur la geometrie des espaces de Riemann*, (Gauthier-Villars, Paris, 1951).

Other valuable works are:

H. Flanders, *Differential Forms*, (Academic Press, New York, 1963),

N. Hicks, *Notes on Differential Geometry*, (Van Nostrand Mathematical Studies #3), (D. Van Nostrand, Princeton, N. J., 1965),

T. Willmore, *Introduction to Differential Geometry*, (Oxford University Press, Oxford, 1959).

CONTENTS

I. INTRODUCTION

Hamiltonian cosmology, the study of cosmological models by means of equations of motion in Hamiltonian form, has begun to receive considerable attention recently. In some ways it is surprising that such a study has come only so lately to the fore, and in some ways it is not. The idea of a Hamiltonian approach to general relativity has been extant since the nineteen-thirties, and it should have been noticed that cosmological models are the simplest, non-static, general relativistic systems known and could have provided a testing ground for Hamiltonian ideas. As we shall point out below, however, there was an emphasis on quantization as the ultimate goal of Hamiltonian formulations, and quantization of a system after restrictions have been imposed classically is a questionable procedure in most cases. With this in mind it would seem reasonable to ignore cosmology until a full quantum theory of gravitation is available. This objection has recently been side-stepped to provide an interesting series of conjectures about quantized universes. The realization that the Hamiltonian approach can be useful in examining classical behavior has also spurred current research.

Historical Background

It might seem facetious to give a historical introduction to a subject which at this time is only three years old, but the preceeding paragraph points out that a discussion of the antecedents of the present work is useful in order to delineate the development of emphasis and philosophy in current work.

Hamiltonian methods applied to gravity seem to go back to Rosenfeld[1], who constructed a quantum-mechanical Hamiltonian for linearized general relativity theory. He made no attempt to construct any type of general canonical approach to general relativity. The first attempt to construct such an approach was due to Bergmann and his collaborators[2] who felt that a more general approach was needed in order to proceed with quantization. Other workers such as Dirac[3], and Pirani and Schild[4] attempted during the nineteen-fifties to build up a Hamiltonian formalism for non-linear field theories, general relativity in particular. Basically these researches were thought of as providing a pathway for quantization, so cosmological questions were not approached.

In the late fifties and early sixties Arnowitt, Deser, and Misner[5] (ADM), in a series of papers constructed a Hamiltonian theory of gravitation which differed slightly from previous theories but which was largely based on their philosophy and partly on their system of notation. ADM had quantization of the gravitational field as their eventual goal, but they were able to use their formalism classically to investigate several problems associated with point particles. No attempt was made to use cosmology as a testing ground for this formalism.

The first to notice that cosmologies provided a simple model in which to demonstrate features of Hamiltonian formulation was DeWitt[6]. Following the lead of the investigators mentioned above he applied the Hamiltonian formulation to the closed ($k=+1$) Friedmann universe. He then quantized this universe as a model for quantum gravitational calculations.

At the time that this work was going on, another line of development which would lead to Hamiltonian cosmology was being pursued by Misner[7]. We shall call this *Lagrangian Cosmology*. In the works cited Misner uses the fact that (essentially) Einstein's G_{oo} can be used as a Lagrangian for the other Einstein equations. Once he had obtained a Lagrangian, he constructed the analogue of the total energy of the system. The formal similarity of this to a Hamiltonian, led him to attempt to apply the ADM techniques to cosmological models. His success in this led to a rapid development of Hamiltonian cosmology and its application to a large variety of cosmological problems, both quantum and classical. Work has been done in this direction by Misner himself[8], Chitre[9], Hughston[10], Jacobs[10], Nutku[11], Ryan[12], and Zapolsky[13]. Lagrangian cosmology, which we shall discuss in Appendix A, continues to be used by Hawking[14], and Matzner[15].

The ADM Formulation

In Section II we discuss the ADM formulation of general relativity in which one reduces the Einstein action for general relativity, $I = \int R\sqrt{-g} \; d^4x$, to the form

$$I = \int [p_j \dot{q}_j - N \, C^o - N_i C^i] \; d^4x \tag{1.1}$$

where q_j are related to the metric components on $t = const.$ hypersurfaces, and where N and N_i are $(-g^{oo})^{-1/2}$ and g_{oi} respectively, and where $C^o = C^o(p_j, q_j)$, and $C^i = C^i(p_j, q_j)$. The N's are to be varied independently, which implies the constraints $C^o = 0$, $C^i = 0$. The ADM procedure involves solving the constraints and choosing certain of the p_j and q_j as coordinates to reduce the action to a form

$$I = \int [p_j \dot{q}_j - H(p_j, q_j)dt] \, d^3x .$$
(1.2)

The Dirac Formulation

The Dirac formulation reduces the action of general relativity to a form similar to (1.1), but prefers to regard the motion as being given by $NC^0 + N_i C^i$ acting as a set of four "Hamiltonians", with the constraints left as a set of supplementary equations to be imposed in conjunction with Hamilton's equations. It is difficult to say whether the ADM formulation and the Dirac formulation lead to the same solution classically or quantum-mechanically in Hamiltonian cosmology, as little cosmological work has been done in the Dirac formulation. We shall, however, be able to point out some differences by means of a bastard formulation (which in a light-hearted moment one might be tempted to call the semi-Dirac method) in which the $C^0 = 0$ constraint is solved, but the three $C^i = 0$ constraints are left as supplementary equations. We shall see that this formulation differs from the full formulation both classically and quantum-mechanically. Because the full Dirac method has been used so little in cosmology, we shall not consider it deeply, but mostly restrict our discussion to the ADM formulation, with occasional discussions of other methods where appropriate.

Homogeneous Cosmologies

While observation tells us that the universe is homogeneous and isotropic and has been for some time, implying that the

Friedmann-Robertson-Walker universes are a good description for it,
our current lack of knowledge about the early stages of the universe
leads us to consider more general models, to see which types could be
fit to the present universe. Misner[16], for instance, assigns this
problem considerable importance with the conjecture that any initial
condition of the universe, be it however anisotropic or inhomogeneous,
will evolve into such a universe as we see today. He has called this
conjecture "chaotic cosmology". Such an idea points to the importance
of considering anisotropic and inhomogeneous cosmological models.

At present homogeneous, anisotropic models are being studied
quite thoroughly, while inhomogeneous universes have generally only
been considered in perturbation. Hamiltonian cosmology has been in-
valuable in the study of anisotropic universes and promises to be
equally useful in the study of more general inhomogeneous cosmologies.
A likely candidate for study as an inhomogeneous cosmology is a universe
proposed by Belinskii and Khalatnikov[17]. In the present work, however,
because Hamiltonian cosmology has not yet been applied to inhomogeneous
universes, we shall restrict ourselves to universes which are homogeneous.

The universes which have received detailed examination by Hamiltonian
methods are the Kantowski-Sachs universe[18], and the class of cosmologies
which have $g_{oo} = 1$, $g_{oi} = 0$ and whose three-space ($t=const.$) sections
are one of the nine, three-dimensional spaces with groups of motions
classified by Bianchi[19]. We shall call these *Bianchi-type universes*
(Bianchi type I through Bianchi type IX). We shall discuss the Kantowski-
Sachs universe in detail in a later section. In our discussion of
Bianchi-type universes we shall write the metrics of these model cosmologies
as

$$ds^2 = - dt^2 + g_{ij}(t)\tilde{\sigma}^i\tilde{\sigma}^j \, , \tag{1.3}$$

where the $\tilde{\sigma}^i$ are three one-forms which obey the relations $d\tilde{\sigma}^i = c^i{}_{jk} \tilde{\sigma}^j \wedge \tilde{\sigma}^k$. The classification of the Bianchi types is by means of the nine distinct possible sets of structure constants, $c^i{}_{jk}$. Table I.1 gives the nine different sets of structure constants.

TABLE I.1

Bianchi type	Structure Constants
I	$c^i{}_{jk} = 0 \quad \forall_{i,j,k}$
II	$c^1{}_{23} = -c^1{}_{32} = 1$
III	$c^2{}_{23} = -c^2{}_{32} = 1$
IV	$c^1{}_{13} = -c^1{}_{31} = c^1{}_{23} = -c^1{}_{32} = c^2{}_{23} = -c^2{}_{32} = 1$
V	$c^1{}_{13} = -c^1{}_{31} = c^2{}_{23} = -c^2{}_{32} = 1$
VI	$c^1{}_{13} = -c^1{}_{31} = 1, \quad c^2{}_{23} = -c^2{}_{32} = h$ $(h \neq 0, 1)$
VII	$c^1{}_{32} = -c^1{}_{23} = c^2{}_{13} = -c^2{}_{31} = 1$ $c^2{}_{23} = -c^2{}_{32} = h \quad (h^2 < 4)$
VIII	$c^1{}_{23} = -c^1{}_{32} = c^3{}_{12} = -c^3{}_{21} = c^2{}_{13} = -c^2{}_{31} = 1$
IX	$c^i{}_{jk} = \varepsilon_{ijk}$

In Section II B we discuss the ADM formulation as applied to the Bianchi types and derive a general canonical form for the Einstein equations of such universes.

In Section III we consider in detail various cosmological models. The most interesting of the Bianchi types are types I, V, and IX which contain as special cases (when $g_{ij}(t) = R(t)\delta_{ij}$) the Friedmann-Robertson-Walker $k = 0, -1, +1$ universes[20] respectively. Types I and IX have been most thoroughly studied of the Bianchi types, and as type I is the simplest we begin by discussing it (after Misner[21]). The Kantowski-Sachs universe turns out to be quite similar to the Bianchi types when cast in canonical form, and we consider it next, following the work of Fishbone[22]. The next most complicated objects are, oddly, the Friedmann-Robertson-Walker universes. The complication lies in the fact that the power of the Hamiltonian method overwhelms the problem of these universes and we are left with auxiliary equations as the only physically meaningful statements we can make.

Next in order of complexity are the type IX universes. We divide these universes into three subcases:

1) Diagonal, in which $g_{ij}(t)$ is a diagonal matrix
2) Symmetric,[23] in which $g_{ij}(t)$ has one off-diagonal term
3) General, in which $g_{ij}(t)$ is a general 3 x 3 matrix

The diagonal case is the basis for the well-known "mixmaster universe" of Misner[24]. The symmetric case has been studied by Ryan[12] and by Oszvath[25] in Hamiltonian formulation. The general case has been studied by Ryan.

The rest of the Bianchi-type universes for the cases in which $g_{ij}(t)$ is a diagonal matrix have been studied by Jacobs and Hughston[26].

We consider their work and present their results.

As we describe the classical motions for each of the homogeneous cosmologies we consider, we discuss quantization of each model. Once we have reduced the equations of motion to canonical form it is not difficult to impose quantum commutation relations on variables and their conjugate momenta. Diagonal Bianchi types, the Kantowski-Sachs universe, and the Friedmann-Robertson-Walker universes have been studied quantum-mechanically in the ADM formulation by Misner[21] and Jacobs and Hughston[26], Fishbone[22], and Nutku[27], respectively. As was pointed out before, DeWitt[6] has also considered the Friedmann-Robertson-Walker universes and a comparison is given between his work and that of Nutku.

The Problem of Matter

In the elaboration of the canonical form for the equations of motion for the cosmological models given above we encounter the fact that in certain of the universes we consider (especially those with non-diagonal $g_{ij}(t)$) the postulated form of the metric is inconsistent with a vacuum solution. For this reason, and because it is customary to consider non-empty universes in any case, we find it necessary to ask how we may include non-zero stress tensors in the Hamiltonian formulation.

In order to add matter to the Einstein equations it is necessary to modify the Einstein action $I = \int R\sqrt{-g}\, d^4x$ to read $I = \int (R\sqrt{-g}+L_M)\, d^4x$, where the Lagrangian density L_M satisfies

$$\delta \int L_M\, d^4x = -8\pi \int T_{\mu\nu}(-g)^{1/2}\, \delta g^{\mu\nu} d^4x \qquad (1.4)$$

Once such a modified action is obtained, it is necessary to break up L_M into terms such as $p_i \dot{q}_i$, and $N\ L_M^o$ and $N_i L_M^i$, the first of these introducing new independent coordinates and the second two quantities which modify the constraints $C^o = 0$, $C^i = 0$ to read $C^{o\,\prime} = C^o + L_M^o = 0$, $C^{i\,\prime} = C^i + L_M^i = 0$.

Such a Lagrangian density L_M exists for electromagnetic fields, and its application to Hamiltonian cosmology is best described in Ref. [26]. It is more usual to postulate that the matter in the universe is fluid, with a stress tensor $T_{\mu\nu} = (\rho + p) u_\mu u_\nu + p g_{\mu\nu}$, where ρ is the energy density and p the pressure, and u_μ is the local fluid velocity. The two most usual fluids considered are dust $(p=0)$ and radiation $(p = \frac{1}{3}\rho)$. Such stress tensors pose a problem, as a Lagrangian density L_M for them which obeys Eq. (1.4) is not known in general (see, however, recent work by Schutz[28]). In the Bianchi-type universes, nevertheless, such an L_M can be constructed for any fluid which has $p = (\gamma-1)\rho$, and this is discussed in Section III C. It is shown there that one can reduce this Lagrangian density to the form $L_M = N\ L_M^o + N_i L_M^i$ so this type of matter serves only to modify the equations $C^o = 0$, $C^i = 0$.

The Hamiltonian Formulation Applied to More Complex Systems

Eventually, once the behavior of the anisotropic, homogeneous cosmologies is understood, it will be necessary to consider inhomogeneous cosmologies. The study of such universes is in a very rudimentary stage, there being only one model which allows large-scale inhomogeneities (as opposed to perturbative ones), that of Belinskii and Khalatnikov[17], which has

been studied only in non-Hamiltonian form. Because there are no good
examples of inhomogeneous universes handled by means of a Hamiltonian
formulation we shall not consider them in this work. We shall, however,
discuss recent, non-cosmological studies of metrics which have inhomogeneous space sections. Kuchar has studied the Einstein-Rosen
cylindrical wave metric[29] in the ADM formulation and has derived its
equation of motion in Hamiltonian form. Nutku and Kobre[30] have recently
extended this to the degenerate problem of the Schwarzschild metric.
We discuss these with the hope that they will point the way toward
methods to handle inhomogeneous cosmologies by Hamiltonian methods.

The Uses of Hamiltonian Cosmology

In Section V we describe various uses to which the Hamiltonian
formulation can be put. The major of these uses is the explication
of complicated motions of the universe. Because the state of a Bianchi-
type universe is given by $g_{ij}(t)$ at any one time, the number of degrees
of freedom of the motion is finite, the state being given by the six
independent components of g_{ij}. Misner[24] has introduced the parametrization
$g_{ij}(t) \propto e^{-2\Omega(t)} e^{2\beta}{}_{ij}$, where Ω is a scalar and β_{ij} a traceless 3×3
matrix. He then takes a coordinate condition $t \to \Omega$, that is, he
chooses Ω as his time coordinate. With this choice, the Hamiltonian
becomes one for β_{ij} as functions of Ω. This reduces the problem to one
of giving the five independent components of β_{ij} as functions of Ω.
There are special cases of β_{ij} which have fewer independent components.
Diagonal β's have only two. This implies that the problem of Bianchi-type

universes reduces to that of the motion of a point, the "universe point", throughout a space of from two to five dimensions under the control of a Hamiltonian derived from the ADM procedure. In Bianchi type I universes, for example, the Hamiltonian is equivalent to that of a relativistic, massless particle in force-free motion in two dimensions, while in Bianchi type IX universes in the diagonal case (mixmaster universe) the motion is equivalent to a particle moving in an expanding potential which has exponential walls and a roughly triangular shape (see Fig. (1.1)). More complicated type IX cases also are shown to reduce to two-dimensional motion under the influence of more complicated potentials which are also time-dependent. The other universes we shall discuss also behave equivalently to particles under the influence of more or less complicated, time-dependent potentials.

In many cases, the steepness of the walls of these potentials will allow us to replace the true walls with infinitely hard ones which move in time. Such a replacement will allow us to give gross features of the time development of model universes by studying their motion during bounces from these walls. This idea was first used by Misner[24] for diagonal type IX universes, and more recently by Ryan[12] (who calls this approach *qualitative cosmology*) for more complicated type IX universes. This approach is perhaps the most important use to which the Hamiltonian formulation can be put classically. The pictorial nature of the solution which it gives allows one to express a solution, which would be extremely complicated if expressed analytically, in the form of a diagram which can be easily interpreted. The usefulness and limitations of this procedure are discussed throughout Section V A.

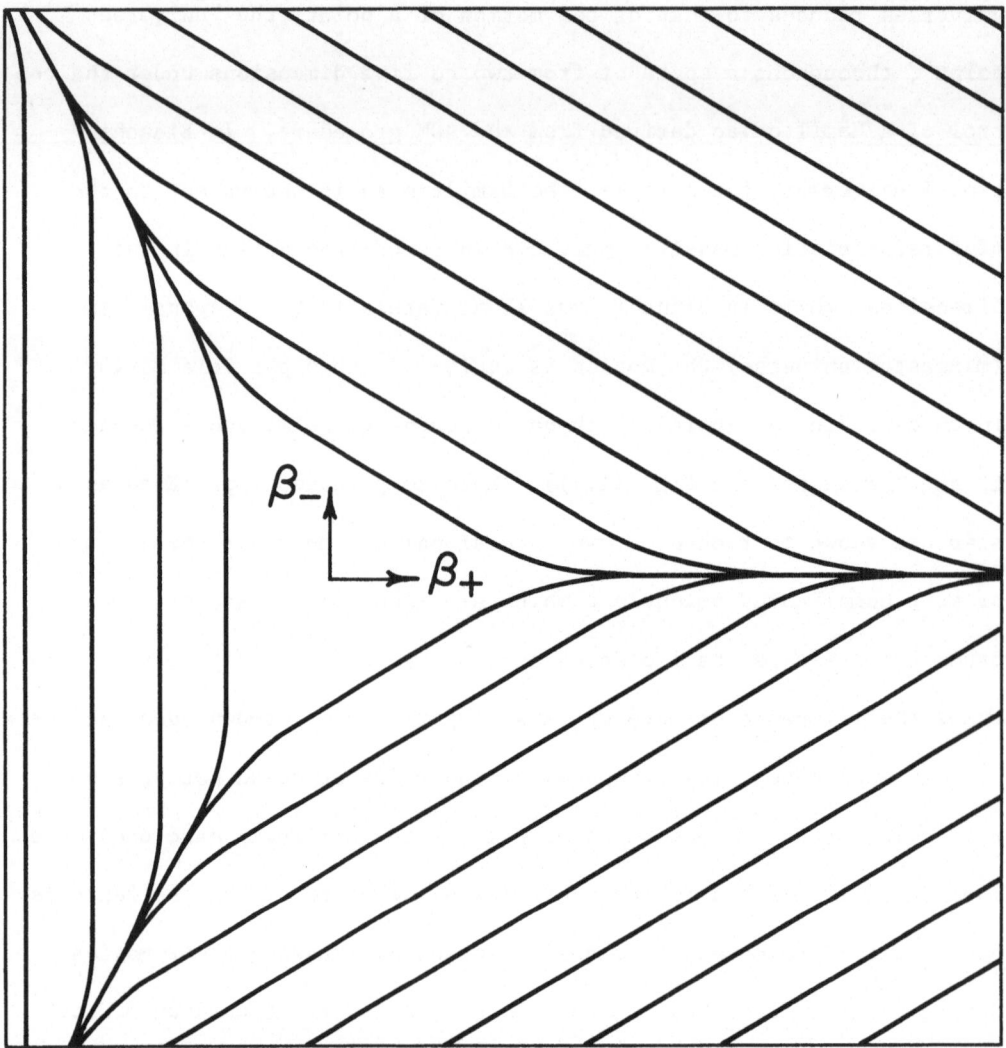

Figure 1.1.   If a diagonal, traceless β-matrix is parameterized by
              $\beta = \mathrm{diag}(\beta_+ + \sqrt{3}\beta_-, \beta_+ - \sqrt{3}\beta_-, -2\beta_+)$, then the potential in
              which the universe point moves is shown in this figure.
              The lines are equipotentials for large $(\beta_+)^2 + (\beta_-)^2$.
              (Figure courtesy C. Misner).

Another of the important uses to which Hamiltonian cosmology
has been put has been the study of the singularities of homogeneous
cosmologies. While the theorems of Hawking et al.[31] have shown that
there must be a singularity for any reasonable cosmology (if quantum
mechanics is ignored), the theorems are cast in such a way that the
character of these singularities is not specifically given. Until the
time when theorems can be given which do this, we must examine each
case separately. The technique of qualitative cosmology has given us
the tool to do this for a large class of homogeneous cosmologies,
and Section V B is devoted to this problem.

For what light it may shed on the problem of the behavior of a
quantized universe near the singularity, the behavior of the quantized
model universes of Section III is considered. In each case we find that
no escape from singularities is provided by quantum mechanics. Of
course, any indication given by such quantized models can only be an
indication and cannot necessarily be taken to imply the behavior of
universes in a fully quantized theory of gravitation.

The final subsection of Section V is devoted to the problem of
"mixing". This is an idea put forth by Misner[24], that in the early
stages of the universe, some sort of disturbance propagating through
the universe could have "smoothed out" any inhomogeneities in the matter,
presenting us with the relatively homogeneous distribution of matter
in the universe which we see today. This idea has been discussed

also by Matzner, Shepley and Warren[32], and Doroshkevich et al.[33].
One of the chief properties of the universe which is needed for mixing
to occur is that of the non-existence of horizons. In the Friedmann
universes, for example, only certain portions of the universe could
be causally connected at the present time. These limits to causal
connectedness are *horizons*. Chitre[9] has shown that for diagonal
Bianchi type IX universes, the behavior of the universe allows the
horizons to disappear in certain directions. Discussion of his work
and its implications forms the bulk of Section V C.

Superspace

Wheeler's concept of superspace[34], the space of all three-geometries,
which is the arena in which the development of the geometries of
general relativity takes place, has played an important role in the
development of Hamiltonian cosmology. This role is discussed in Section
VI. The three-spaces of Bianchi-type cosmologies depend on from two
to five parameters, so the subspace of superspace inhabited by them
has from two to five dimensions. This makes these cosmologies valuable
probes with which to study conjectures relating to superspace as a whole.

One program which has been of interest recently has been the attempt
to put a metric on superspace. This has been considered mainly by
Misner[35], DeWitt[36], and Gowdy[37]. This work is discussed.

Quantization

In Section VII we consider the problem of quantized cosmology, bringing together all the threads of discussion which have run through the other sections. After reinforcing the caveat given above that quantized models cannot be blithely assumed to represent the cosmological limit of quantized general relativity, we discuss problems encountered in quantizing the Bianchi types. Because the Hamiltonians for the Bianchi types are all square-root Hamiltonians, we discuss methods of handling them. Methods for dealing with square-root Hamiltonians are well known from ordinary relativistic quantum mechanics, and we compare the best-known of these.

More serious problems come in due to the different possible factor-orderings in the quantum-mechanical Hamiltonian, and to different ordering of solutions of the constraints $C^i = 0$ with quantization. These problems are discussed, and pros and cons of each method are adduced and examined.

II. THE ADM FORMALISM APPLIED TO HOMOGENEOUS COSMOLOGIES

A. The ADM Hamiltonian Formalism

The ADM formalism[5] was developed with an eye toward quantizing the gravitational field by recasting the Einstein equations in canonical form. In order to achieve this it was necessary to do three things: 1) Rewrite the Einstein equations in a first-order form; 2) Deal with the fact, which becomes evident from this, that the Lagrangian for gravity is in an "already parameterized" form, that is, deal with the invariance of the graviational field under general coordinate transformations; and 3) Reduce the Lagrangian to canonical form.

The ADM procedure for achieving these ends begins by rewriting the Einstein Lagrangian in the Palatini form.[38] Next one re-parameterizes the quantities ($\Gamma^{\alpha}_{\mu\nu}$ and $g_{\mu\nu}$) which appear in this form of the Lagrangian. This re-parameterization introduces the quantities $N = (^{4}\text{-}g^{00})^{-1/2}$, $N_i = {}^{4}g_{oi}$, π_{ij}, c^{o} and c^{i}, where the superscript ^{4} denotes a four dimensional quantity, the superscript ^{3} a three dimensional quantity. The g_{ij} are the components of the metric on $t = constant$ hypersurfaces ($g_{ij} \equiv {}^{4}g_{ij}$), and the π^{ij} are quantities defined in terms of the $^{4}g_{\mu\nu}$ and $\Gamma^{\mu}_{\nu\alpha}$. The quantities c^{o} and c^{i} are defined in terms of π^{ij} and g_{ij} and their space derivatives. ADM show that the Einstein action $I = \int (-g)^{1/2} R_{\mu\nu} g^{\mu\nu} d^{4}x$ reduces to

$$I = \int [\pi^{ij} (\frac{\partial g_{ij}}{\partial t}) - Nc^{o} - N_i c^{i}] d^{4}x \qquad (2.1)$$

(where we have discarded a total divergence). The Einstein equations are given in Ref. [5] and consist of two sets of equations, one for $\partial g_{ij}/\partial t$

and the other for $\frac{\partial \pi_{ij}}{\partial t}$, both involving π_{ij}, g_{ij} and their space

derivatives and N and N_i and their space derivatives. We also have

the equations obtained by varying N and N_i, $C^o = 0$, $C^i = 0$. The

derivative equations are the time-step equations which move g_{ij} and

π^{ij} from one $t = constant$ surface to another. The C^o, $C^i = 0$ equations

are a set of constraints on π^{ij} and g_{ij} that keep them from being

specified arbitrarily at any one time.

Up to this point the reduction of the **Einstein** equations is well

known and parallels the work of earlier investigators. The heart of

the ADM procedure is their reduction of this Lagrangian to canonical

form. This is accomplished by a two-step process. The first step is

to solve $C^o = 0$, $C^i = 0$ for **four** of the g^{ij} and π^{ij} (which number up to

twelve, depending on the symmetries of the metric) and insert these

solutions into the **action** (2.1). The next step is to choose four

coordinate conditions which reduce the number of independent variables in

the Lagrangian to four, roughly speaking to two of the g_{ij} and their

conjugate π's. The two steps just outlined are rather broad in their

prescription, and justifiably so. The two choices, that is, which

variables to eliminate and what coordinate conditions to choose must be

made in concert and made adroitly to obtain the final action in canon-

ical form.

ADM found that choices of functions of the g_{ij} and π^{ij} as coordinates

led most naturally to canonical form for the action (2.1). More **recently**,

in applications to cosmology it has been more useful to choose some

coordinates as functions of the g_{ij} and π^{ij} and to choose the rest by

means of specific choices of N and N_i. These two ways of choosing coordinate conditions are, of course, entirely equivalent, a choice of N and N_i fixing the coordinate system, and a choice of coordinates uniquely determining N and N_i by the requirement that the variational system be equivalent to the Einstein equations. The mixing of these two ways of choosing coordinates has pitfalls despite its usefulness. It is often tempting to overdetermine the problem by making coordinate choices and choosing N and N_i to fit some preconceived criterion (such as choosing $N_i = 0$ in a cosmological problem). The ADM procedure tends to mask the contradictions in these choices by giving a perfectly con- sistent set of canonical equations. The problem is that these equations are *not* equivalent to Einstein's equations. Therefore, whenever one mixes his methods of choosing coordinate conditions it is necessary to exercise **great** care that these choices are, in fact, consistent.

In applying the ADM approach to cosmology, it has been found useful in some cases to avoid carrying out the procedure completely and to leave the substitution of solutions of the space constraints into the action and arrive at a constrained Hamiltonian version of the Einstein equations. When and why this is useful will be discussed later.

We note here that, of course, there is no rule for choosing which coordinates we shall use, indeed more than one such choice can be made. This leaves the discussion of the general method rather vague, because the choices depend on the goal one is trying to achieve. In outline, however, the procedure is as follows:

1) Compute the action and the constraints

$$I = \int \left[(\pi^{ij} \frac{\partial g_{ij}}{\partial t}) - NC^0 - N_i C^i \right] d^4x \tag{2.2}$$

$$C^0 = -(g)^{1/2} \{ {}^3R + g^{-1} [\frac{1}{2}(\pi^k{}_k)^2 - \pi^{ij}\pi_{ij}] \} \tag{2.3}$$

$$C^i = -2\pi^{ij}{}_{|j} \tag{2.4}$$

where ${}^3R$ is the scalar curvature of the $t = constant$ surfaces.

2) By examining the form of I, decide which choice of functions of g_{ij} and π^{ij} as coordinates will lead to the most useful canonical form, and solve the constraints to eliminate such variables as are needed to put I in canonical form.

3) If one's coordinate choices have been made by a combination of choosing functions of the g_{ij} and π^{ij} and choices of N and N^i, check to see that this is consistent by examining the equations for $\frac{\partial g_{ij}}{\partial t}$ and $\frac{\partial \pi^{ij}}{\partial t}$.

4) If it is more convenient, leave the $C^i = 0$ equations as a constraint for the final Hamiltonian.

B. Homogeneous Cosmologies

1. Discussion

The ADM procedure is very aptly applied to homogeneous cosmologies in many cases. If we write the metric of any homogeneous cosmology with a cosmic time as

$$- dt^2 + g_{ij}(x, t) dx^i dx^j , \tag{2.5}$$

then the total volume of the universes at any one time is $V = \int \sqrt{g} \, d^3x$ (where g is $\det(g_{ij})$), where V is finite if the universe is "closed" and infinite otherwise. If we artificially close "open" universes by

restricting the ranges of the coordinates, V is a function of time only.
If we can, as we can in many cases, factor V out of the g_{ij}, this "volume"
provides a measure of the time if V is a monotonic function of t. In
many well-known, homogeneous cosmologies V is easily factorable, (*vide*
the Friedmann universes) and is usually well behaved enough so that
regions in which it is a monotonic function of t are easily identified.
In each of these regions V is a good time variable and we might hope to
take it as the function of the g_{ij} and π^{ij} to be the intrinsic "time"
in our canonical form of the action for this metric. Because of the
homogeneity of the space sections, we might expect that the final form
of the action could have minimal variables which would have no space
dependence, that is, the metric would be given by objects, numbering
zero to two which would be functions of time (or of V) only. In the
cases in which these are one or two determining parameters of the metric
the aptness of the ADM approach becomes most evident. The problem reduces
in this case, to one equivalent to the motion of a single particle in a
space of one or two dimensions. Solutions and techniques are readily
available to handle such problems. In cases where we leave some of the
constraints unsolved, the problem is still equivalent to that of a
particle, but the dimension of the space in which the equivalent particle
moves increases from three to five. The remaining case, when the number of
independent objects labeling the metric after we impose our coordinate
conditions is zero, occurs when the symmetry of the space considered
allows the constraints to exhaust the dynamical content of the system.
Handling such problems, especially when quantization is considered, is
more difficult than handling the seemingly more complicated problems

in which there is some dynamical content left when the constraints have been solved. The usual Friedmann solutions are an example of such a case.

A class of homogeneous cosmologies in which V is easily factored out, and for which it is not difficult to find a set of variables which describe the metric completely and which do not depend on the space co-ordinates, are the Bianchi-type universes. While these do not exhaust all the possible homogeneous cosmologies, they are a large class and their study is quite rewarding. We shall proceed to apply the ADM formalism to them.

2. Bianchi-type Universes

From Section I we write

$$ds^2 = -dt^2 + g_{ij}(t) \, \tilde{\sigma}^i \, \tilde{\sigma}^j, \tag{2.6}$$

and parametrize g_{ij} by means of Misner's[24] parametrization, $g_{ij} = R_o^2 e^{-2\Omega} e^{2\beta}{}_{ij}$. In general we want to reparametrize β_{ij} by means of para-meters numbering from two to five, β_A, chosen for convenience in carrying out our procedure, in order to avoid a plethora of indices. If we re-write the action in terms of β_{ij}, Ω and π^{ij} we find

$$I = (16\pi)^{-1} \int 2[-\pi^k{}_k \frac{d\Omega}{dt} + (e^\beta \pi^* e^{-\beta})_{ij} \frac{d\beta_{ij}}{d\Omega} \frac{d\Omega}{dt}]d^4x \tag{2.7}$$

We now want functions of Ω, β_{ij} and π^{ij} to serve as coordinates. If we choose Ω as our time variable we find that

$$I = (16\pi)^{-1} \int 2[(e^\beta \pi^* e^{-\beta})_{ij} \, d\beta_{ij} - (\pi^k{}_k)d\Omega]d^3x. \tag{2.8}$$

We expect the π^{ij} to be homogeneous since they are conjugate to the β_{ij} which are homogeneous, so we can integrate over the space variables.

In the orthonormal frame $(d\Omega, \tilde{\sigma}^1, \tilde{\sigma}^2, \tilde{\sigma}^3)$, $d^3x = \tilde{\sigma}^1 \wedge \tilde{\sigma}^2 \wedge \tilde{\sigma}^3$ and if we integrate over the $\tilde{\sigma}^i$ for closed spaces and artificially close open spaces and integrate we get $\int \tilde{\sigma}^1 \wedge \tilde{\sigma}^2 \wedge \tilde{\sigma}^3 = const$. Because for type IX universes this constant is $(4\pi)^2$, Misner[21] chooses this condition for type I universes which are open. We shall make this choice in general (rescaling the forms and the C^i_{jk} if necessary). This implies

$$I = (2\pi) \int [(e^\beta \pi^* \star e^{-\beta})_{ij} \, d\beta_{ij} - (\pi^k_{\ k}) d\Omega]. \tag{2.9}$$

This action will be in canonical form if we can find a set of parameters β_A for β_{ij} and a matrix p_{ij} (parametrized by a set of parameters p^A) such that $(2\pi) (e^\beta \pi^* \star e^{-\beta})_{ij} \, d\beta_{ij} = p_{ij} d\beta_{ij} = p^A d\beta_A$, **and if we can solve** the constraints for $\pi^k_{\ k}$ in terms of p_{ij}, β_{ij} and Ω. If we examine the constraint $C^0 = 0$ we find that the only other terms appearing in the equation besides $\pi^k_{\ k}$ are $\pi^{ij} \pi_{ij}$, ^3R, and g. In our frame $g = R^6_0 e^{-6\Omega}$. After Ryan[12] we define $p_{ij} = 2\pi (e^\beta \pi^* \star e^{-\beta})_{ij} - \frac{2\pi}{3} \delta_{ij} \pi^\ell_{\ \ell}$ (this definition, while more complicated then $p_{ij} = (e^\beta \pi^* \star e^{-\beta})_{ij}$, leads to a major simplification later). This implies $(4\pi^2) \pi^{ij} \pi_{ij} - \frac{4\pi^2}{3} (\pi^\ell_{\ \ell})^2 = p_{ij} p_{ij}$. The scalar curvature of the $t = const.$ surfaces, ^3R, by the definition of homogeneity, must be an algebraic function of β_{ij} and Ω. Thus, we find our action reduced to

$$I = 2\pi \int [p^A \, d\beta_A - H(p^A, \beta_A, \Omega) d\Omega] \quad . \tag{2.10}$$

We now need such a set of parameters for β_{ij} and p_{ij}. A useful parameterization for β_{ij} has been given by Ryan[12]. It takes advantage of the fact that any real symmetric matrix A with non-zero determinant can be written as

$A = R^{-1} A^d R$, where A^d is a real, diagonal matrix and R is a rotation

matrix. We have an excellent parameterization for rotation matrices in

the Euler matrices; in fact we can write $\begin{bmatrix} cos\phi sin\phi 0 \\ -sin\phi cos\phi 0 \\ 0 \quad 0 \quad 1 \end{bmatrix} = e^{\phi \kappa_3}$ where

$\kappa_3 = \begin{bmatrix} 0 & 1 & 0 \\ -1 & 0 & 0 \\ 0 & 0 & 0 \end{bmatrix}$ and $\begin{bmatrix} 1 & 0 & 0 \\ 0 & cos\theta sin\theta \\ 0 & -sin\theta cos\theta \end{bmatrix} = e^{\theta \kappa_1}$, where $\kappa_1 = \begin{bmatrix} 0 & 0 & 0 \\ 0 & 0 & 1 \\ 0 & -1 & 0 \end{bmatrix}$,

which leads to

$$R_{\phi\theta\psi} = e^{\psi\kappa_3} e^{\theta\kappa_1} e^{\phi\kappa_3} \qquad . \qquad (2.11)$$

With this any β_{ij} can be written as

$$\beta = e^{-\psi\kappa_3} e^{-\theta\kappa_1} e^{-\phi\kappa_3} \beta_d e^{\phi\kappa_3} e^{\theta\kappa_1} e^{\psi\kappa_3} , \qquad (2.12)$$

where β_d is a diagonal, traceless matrix. Misner[24] parametrizes a diag-

onal, traceless matrix by $\beta_d = diag(\beta_+ + \sqrt{3}\beta_-, \beta_+ - \sqrt{3}\beta_-, -2\beta_+)$. The calculation

of $d\beta_{ij}$ for (2.12) is given by Ryan.[12] We find

$$d\beta = R^{-1} \{\alpha_1 \, d\beta_+ + \alpha_2 \, d\beta_- + \alpha_3 \, \tilde{\sigma}^3 \, sinh(2\sqrt{3} \, \beta_-) - \alpha_4 \, \tilde{\sigma}^1 \, sinh(3\beta_+ + \sqrt{3} \, \beta_-)$$

$$-\alpha_5 \, \tilde{\sigma}^2 \, sinh(3\beta_+ - \sqrt{3} \, \beta_-)\}R , \qquad (2.13)$$

where the set of matrices $\alpha_1----\alpha_5$ is $\alpha_1 = diag(1, 1, -2)$,

$\alpha_2 = diag(\sqrt{3} , -\sqrt{3} , 0)$ $\alpha_3 = \begin{bmatrix} 0 & 1 & 0 \\ 1 & 0 & 0 \\ 0 & 0 & 0 \end{bmatrix}$ $\alpha_4 = \begin{bmatrix} 0 & 0 & 1 \\ 0 & 0 & 0 \\ 1 & 0 & 0 \end{bmatrix}$

$\alpha_5 = \begin{bmatrix} 0 & 0 & 0 \\ 0 & 0 & 1 \\ 0 & 1 & 0 \end{bmatrix}$, and the $\tilde{\sigma}^i$ are three differential one-forms, invariant

on the three sphere. They are

$$\tilde{\sigma}^1 = sin\phi d\theta - cos\phi sin\theta d\psi$$

$$\tilde{\sigma}^2 = cos\phi d\theta + sin\phi sin\theta d\psi \qquad (2.14)$$

$$\tilde{\sigma}^3 = -(d\phi + cos\theta d\psi).$$

It is useful to note that the α-matrices are linearly independent, so they form a basis for the set of traceless, symmetric 3 × 3 matrices. If we define an inner product of two of these matrices as the trace of their matrix product, they form an orthogonal basis. With this in mind, computation of p_{ij} is not difficult. We want a traceless, symmetric matrix, so we can write

$$p_{ij} = R^{-1} \, p_K \, \alpha_K \, R \qquad (2.15)$$

and we want the inner product of p and $d\beta$ to be $p_+ \, d\beta_+ + p_- \, d\beta_- + p_\psi \, d\psi + p_\phi \, d\phi + p_\theta \, d\theta$ if we parametrize p_{ij} by p_+, p_-, p_ψ, p_ϕ, and p_θ . The orthogonality of the α_i gives us

$$
6p_{ij} = R^{-1} \left\{ \alpha_1 \, p_+ + \alpha_2 \, p_- - \alpha_3 \, \frac{3p_\phi}{sinh(2\sqrt{3}\,\beta_-)} \right.
$$

$$
- \alpha_4 \, \frac{3(p_\psi sin\phi - p_\phi cos\theta sin\phi + p_\theta cos\phi sin\theta)}{sin\theta sinh(3\,\beta_+ + \sqrt{3}\,\beta_-)} \qquad (2.16)
$$

$$
\left. - \alpha_5 \, \frac{3(p_\theta sin^2\phi sin\theta - p_\psi sin\phi cos\phi + p_\phi cos\phi sin\phi cos\theta)}{sin\phi sin\theta sinh(3\,\beta_+ - \sqrt{3}\,\beta_-)} \right\} R
$$

We have put off discussing the three $c^i = 0$ constraints to this point because their computation depends on the structure of the $t = constant$ hypersurfaces, that is, on which Bianchi type we are considering. However, because the π^{ij} are homogeneous, we can say that the c^i will be algebraic functions of π^{ij}, β_{ij} and Ω. By our definition of π^{ij}, $\pi^{ij} = \pi^{ij}(p^A, \beta_A, \Omega, \pi^l{}_l)$, and as we are using c^0 to eliminate $\pi^l{}_l$ as a function of p_A, β_A and Ω we find $c^i = c^i(p^A, \beta_A, \Omega)$. Thus for *all*

Bianchi types the ADM procedure leads to

$$I = \int \ [\ p^A \ d\beta_A - H(p^A, \ \beta_A, \ \Omega)d\Omega\ ], \tag{2.17}$$

with constraints

$$c^i(p^A, \ \beta_A, \ \Omega) \ = \ 0 \tag{2.18}$$

and a Hamiltonian

$$H^2 = (2\pi)^2(\pi^\ell_\ell)^2 = 6tr[(p)^2] - 24\pi^2 \ g^3R \ . \tag{2.19}$$

Note that H is a *square-root Hamiltonian*.

We can now solve the c^i constraints and substitute the solutions into the action, or we can simply leave them as constraints and deal with a constrained Hamiltonian. We shall discuss these two possibilities as we discuss the problem of coordinates.

We have made one coordinate choice so far, that of Ω as our time variable. Since the space coordinates do not appear, their choice has not been crucial, and the possibility of their choice is an important freedom left to us. We can make these choices in many ways. Misner[21] has shown for types I and IX that the equation $g^{ij}\frac{\delta I}{\delta \pi^{ij}} = 0$ derived from the action (2.1) implies $N=H^{-1}e^{-3\Omega}(12\pi R_o^3)$. This relations holds for all Bianchi types and for our Ω-time. The space coordinates then are best chosen by choosing the N_i. In order to see this, let us study the structure of the full action now that we have inserted our parametrization.

We have

$$
\begin{aligned}
I = \int \big[& p_+ \, d\beta_+ + p_- \, d\beta_- + p_\phi \, d\phi + p_\psi \, d\psi + p_\theta \, d\theta \\
& - H(\Omega, p_\pm, \, \beta_\pm, \, \phi, \, p_\phi, \, \psi, \, p_\psi, \, \theta, \, p_\theta) d\Omega \\
& - N_i \, C^i (\Omega, p_\pm, \beta_\pm, \, \phi, \, p_\phi, \, \psi, \, p_\psi, \, \theta, \, p_\theta) \big].
\end{aligned}
\tag{2.20}
$$

We can now do one of three things. We can choose the coordinate condition $N_i = 0$ which leads to a completely Hamiltonian system for $\beta_\pm$, ϕ, ψ, θ. (but with three ancillary conditions, $C^i = 0$ to serve as constraints which do not derive from the reduced action). We call this first possibility the *All-Hamiltonian system*. The second possibility is to solve the constraints $C^i = 0$, and plug the solutions back into H. This leads to the minimal number of variables needed to describe the system, and again we have ancillary equations which describe the variables which do not appear in the reduced action in terms of the others. We call this the *minimal system*. The final method is to vary the entire action (2.20) with no **restrictions**. This gives us the ancillary conditions which are needed for the other two cases but which do not follow from their respective variational principles. Table II.1 gives a schematic **representation** of all these cases.

Notice that in the minimal system we still have the freedom of choice of N_i as functions of Ω. These can be chosen to be zero as in the all-Hamiltonian case, causing the eliminated variables to mimic the behavior of the same variables in that case. We can, however, choose the N_i in such a way as to make the eliminated variables behave in any way we wish, in particular we could choose the N_i to make them zero for all time.

TABLE II.1

$$I = \int L(p_\pm, \beta_\pm, p_\phi, \phi, p_\psi, \psi, p_\theta, \theta, N_i(\Omega))$$

Complete Variational System

$$\delta p_\psi :: \dot\psi = \frac{\partial H}{\partial p_\psi} + N_i \frac{\partial C_i'}{\partial p_\psi}$$

$$\delta \psi :: \dot p_\psi = -\frac{\partial H}{\partial \psi} - N_i \frac{\partial C_i'}{\partial \psi}$$

$$\delta p_\phi :: \dot\phi = \frac{\partial H}{\partial p_\phi} + N_i \frac{\partial C_i'}{\partial p_\phi}$$

$$\delta \phi :: \dot p_\phi = -\frac{\partial H}{\partial \phi} - N_i \frac{\partial C_i'}{\partial \phi}$$

$$\delta N_i :: C_i' = 0$$

$$\delta p_\pm :: \dot\beta_\pm = \frac{\partial H}{\partial p_\pm} + N_i \frac{\partial C_i'}{\partial p_\pm}$$

$$\delta \beta_\pm :: \dot p_\pm = -\frac{\partial H}{\partial \beta_\pm} - N_i \frac{\partial C_i'}{\partial \beta_\pm}$$

Constrained System
Solve $C_i' = 0$

$$\dot\psi = \frac{\partial H}{\partial p_\psi}\Big|_{C_i'=0}$$

$$\dot\theta = \frac{\partial H}{\partial p_\theta}\Big|_{C_i'=0}$$

$$\dot p_\theta = -\frac{\partial H}{\partial \theta}\Big|_{C_i'=0}$$

$$\dot\phi = \frac{\partial H}{\partial p_\phi}\Big|_{C_i'=0}$$

$$\dot p_\phi = -\frac{\partial H}{\partial \phi}\Big|_{C_i'=0}$$

$$\dot\beta_\pm = \frac{\partial H}{\partial p_\pm}\Big|_{C_i'=0}$$

$$\dot p_\pm = -\frac{\partial H}{\partial \beta_\pm}\Big|_{C_i'=0}$$

All Hamiltonian System
$(N_i = 0)$

$$\dot\psi = \frac{\partial H}{\partial p_\psi}$$

$$\dot\theta = \frac{\partial H}{\partial p_\theta}$$

$$\dot p_\theta = -\frac{\partial H}{\partial \theta}$$

$$\dot\phi = \frac{\partial H}{\partial p_\phi}$$

$$\dot p_\phi = -\frac{\partial H}{\partial \phi}$$

$$\dot\beta_\pm = \frac{\partial H}{\partial p_\pm}$$

$$\dot p_\pm = -\frac{\partial H}{\partial \beta_\pm}$$

TABLE II.1 (Continued)

$$I = \int L(p_\pm, \beta_\pm, p_\phi, \phi, p_\psi, \psi, p_\theta, \theta, N_i(\Omega))$$

All Hamiltonian System
$(N_i = 0)$

$$\dot{p}_\psi = -\frac{\partial H}{\partial \psi}$$

Complete Variational System

$$\delta\theta : \dot{p}_\theta = -\frac{\partial H}{\partial \theta} - N_i \frac{\partial C'_i}{\partial \theta}$$

$$\delta p_\theta : \dot{\theta} = \frac{\partial H}{\partial p_\theta} + N_i \frac{\partial C'_i}{\partial p_\theta}$$

Constrained System
Solve $C'_i = 0$

$$\dot{p}_\psi = -\frac{\partial H}{\partial \psi}\Big|_{C'_i=0}$$

Ancillary equations:

$C'_i = 0$

Ancillary equations:

$C'_i = 0$ will eliminate three variables, so we shall need three equations from the complete system to determine them, with N_i inserted as an arbitrary function of Ω.

* In this column we mean $\frac{\partial H}{\partial x}\Big|_{C'_i=0}$, if x is an eliminated coordinate or momentum, to contain only algebraic combinations of non-eliminated variables, where for eliminated variables we mean for $\frac{\partial H}{\partial x}\Big|_{C'_i=0}$ to be $\frac{\partial}{\partial x}(H|_{C'_i=0})$.

Non-zero choices of N_i, however, introduce g_{oi} terms into the metric, which leads to difficulties in interpretation as serious as those encountered if the variables were left in the problem.

We can now begin to ask what the differences between the various Bianchi types will be. The general procedure just outlined leaves $c^i(\Omega, p_\pm, \beta_\pm, p_\phi, \phi, p_\psi, \psi, p_\theta, \theta)$ and $^3R(\Omega, \beta_\pm, \phi, \psi, \theta)$ to be calculated for the various Bianchi types. These four quantities are *all* that need to be calculated to reduce any particular Bianchi type to canonical form. With our general form at hand, we can begin to consider possible special cases in which we restrict the metric by imposing symmetries. The most well-known examples are types I, V, IX with the choice $\beta_+ = \beta_- = \phi = \psi = \theta = 0$. These are the $k = 0, -1, +1$ Friedmann universes, respectively. Another set of restriction considered by Misner[21], and Jacobs and Hughston[26] for various types has been the choice of $\beta_\pm \neq 0$, $\phi = \psi = \theta = 0$, that is, the β-matrix is chosen to be diagonal. Ryan[12] has considered a type IX case in which $\beta_\pm \neq 0$, $\phi \neq 0$, $\psi = \theta = 0$. The three possible type IX cases provide an example of one pitfall in the choice of special cases. In any such choice we must take care that the Einstein equations allow such a choice; that is, that as the metric develops in time it retains the form we have chosen for it. This problem occurs, for example, in Type IX universes. While the choices $\beta_\pm = 0$, $\phi = \psi = \theta = 0$ (Friedmann $k = +1$), $\beta_\pm \neq 0, \phi \neq 0$, $\psi = \theta = 0$, $\beta_\pm \neq 0$, $\phi \neq$ $\psi \neq \theta \neq 0$ are possible and consistent, if we choose $\beta_\pm \neq 0$, $\phi \neq 0$, $\theta \neq 0$, $\psi = 0$ at any one time, this universe will develop into one with all variables non-zero.

C. The Problem of Matter in the ADM Formalism

The above discussion about the necessity of choices of restrictions on universes being consistent leads naturally to another problem of consistency. We may postulate a complicated, empty, Bianchi-type universe, but we cannot be sure that such an empty universe is possible, that is, the constraints may show us that the number of degrees of freedom we have postulated is too large to be satisfied by a simple vacuum stress tensor. Such a case occurs in Bianchi type IX universes. In the case where $\psi = \theta = 0$, $\phi \neq 0$, one of the space constraints reads $p_\phi = 0$ for a zero stress tensor. This implies that $\phi = const$, or by a proper choice, $\phi = 0$ which reduces us to the case in which β is a diagonal metric. Because the $c^i = 0$ constraints are the R_{oi} Einstein equations, $R_{oi} = 0$ in this case implies that β is diagonal. If, however, we were to introduce pressureless fluid matter into the problem, whose local fluid velocity has non-zero space components, then we would have $R_{oi} = 8\pi\rho u_o u_i$ which could give a soluble problem.

The need for matter terms in the Einstein equations leads us to ask how we may introduce matter into the ADM formalism. In order to do this we need a Lagrangian for the type of matter or energy we wish to consider, that is we need a scalar density L_M such that

$$\delta \int L_M \, d^4x = -8\pi \int T_{\mu\nu} (-{}^4g)^{1/2} \, \delta g^{\mu\nu} \, d^4x \,. \tag{2.21}$$

In the case of electromagnetic energy and the various matter fields of quantum field theory, general relativistic Lagrangians are well known, and by rewriting them in terms of N, N_i and g_{ij} and including terms which modify the constraints (proportional to N and N_i), terms which solely

determine the motion of the matter (involving neither N and N_i or g_{ij}),

and interaction terms (multiplicative combinations of g_{ij} and the matter

quantities). Once the new degrees of freedom have been identified, we

may attempt to proceed with our reduction essentially as before.

Traditionally, however, cosmologists have tended to introduce fluid

stress tensors into cosmological problems. This presents a difficulty

in Hamiltonian cosmology because the study of Lagrangians for fluid mat-

ter has not proceeded far enough to give such a Lagrangian in an

immediately useful form . There are cases, however, where a Lagrangian

may be constructed to fit a particular problem. One such case occurs

in non-diagonal type IX universes. One of these type IX universes also

happens to illustrate another possible circumstance. The fluid equations

of motion may be soluble in terms of the gravitational degrees of

freedom and constants of motion. In this case the fluid introduces no

new degrees of freedom and serves only to modify the constraints, that

is $C^O \rightarrow C'^O$, $C^i \rightarrow C'^i$, where C'^O, C'^i are functions of Ω, $p_\pm$, $\beta_\pm$, p_ϕ

p_ψ p_θ, ϕ, ψ, θ. This occurs in type IX universes in which $\phi \neq 0$,

$\psi = \theta = 0$.

We cannot say more about the inclusion of matter terms because of

the need to consider each type of universe individually. We shall

discuss matter terms in examples given below.

D. Problems in Quantizing Bianchi-type Universes

We point out here the major problems in quantizing Bianchi-type

universes. Two problems which stand out are: 1) The fact that the

Hamiltonian for these universes is a square-root Hamiltonian

2) The Hamiltonian is explicitly time-dependent.

The first of these problems can be solved in a variety of ways. In fact there is an *embarras de richesses* in methods of handling square-root Hamiltonians, familiar from the quantization of relativistic particle mechanics. The three methods we shall consider are: 1) The square-root method of Schweber et al.[38], in which we use Fourier analysis to obtain a square root of the Hamiltonian operator; 2) The Dirac method in which we linearize the square-root Hamiltonian in terms of matrices; 3) The Schrödinger -Klein-Gordon (SKG)[39] method in which we use the squared Hamiltonian to obtain a second-order equation. Each of these three methods has strengths and weaknesses when applied to Bianchi-type cosmologies. We shall discuss them in context when we consider specific models.

The fact that the Hamiltonian is explicitly time-dependent is theoretically no difficulty, but examples of such Hamiltonians are rare in well known quantum-mechanical problems and techniques for handling them are not well developed. This practical difficulty can be surmounted in several ways and we shall discuss these ways below.

A third problem in quantization shows up when we find it more convenient not to solve the constraints completely, but leave a constrained Hamiltonian. In such a case we must cope with the difficult problem of quantizing a constrained system.

The final problem we shall consider is that of matter. When it is necessary to add a matter term to the classical equations to obtain a solution, how can we treat the quantum problem? Of course, a vacuum quantum solution may exist, even when no corresponding classical solution exists.   There may be, however, situations when no vacuum quantum

solution exists, or when we need to consider matter-filled universes

for other reasons.   In such cases we must be able to quantize the inter-

acting system of gravitational field and matter.   We shall discuss this

further below.

III. THE HAMILTONIAN FORMALISM:
SIMPLE EXAMPLES, THEIR CLASSICAL AND QUANTUM BEHAVIOR

A. Bianchi Type I Universes

1. *Classical Behavior*

The cosmological models for which we obtain perhaps the simplest dynamical Hamiltonian are the Bianchi type I universes with β a diagonal matrix. In these the structure constants are all zero and the $\tilde{\sigma}^i$ reduce to dx^i, coordinate differentials. In empty type I universes, the constraints $\pi^{ij}{}_{|j}$ are identically satisfied, so to use our general Hamiltonian (2.19) **we need only insert the conditions** $\theta = \psi = \phi = p_\theta = p_\psi = p_\phi = 0$ into it, and compute ${}^3R$. Obviously ${}^3R = 0$. From this we have

$$H^2 = p_+{}^2 + p_-{}^2 \ . \tag{3.1}$$

Misner[21] was the first to write the Hamiltonian for Bianchi type I universes in this form. Bianchi type I cosmologies have been investigated very thoroughly from many points of view[40], but the simplicity of the Hamiltonian formulation makes the behavior of these universes much easier to understand.

We see from the form of H that the problem of the time development of the universe is equivalent to the free motion in two dimensions of a massless, relativistic particle. Because β can be set equal to zero at any one time, we need only consider trajectories (in the β_+, β_--plane) through $\beta_+ = \beta_- = 0$. It is obvious that these trajectories are straight lines, and Hamilton's equations imply that $p_\pm$ and H are constants of the motion. The equations of motion for $\beta_\pm$ are

$$\dot{\beta}_{\pm} = \frac{p_{\pm}}{H} \; , \tag{3.2}$$

so $(\dot{\beta}_+)^2 + (\dot{\beta}_-)^2 = 1$. This means that the universe point moves with unit Ω velocity (in the sense of $d\beta/d\Omega$) along the trajectories. The general solution is then

$$\beta_+ = \Omega \; \cos\theta, \tag{3.3}$$

$$\beta_- = \Omega \; \sin\theta \; ,$$

where θ is a constant. To complete our description of the motion of the universe we can write

$$ds^2 = -N^2 d\Omega^2 + R_o^2 \; e^{-2\Omega} (e^{2\beta})_{ij} \; dx^i \; dx^j, \tag{3.4}$$

with N a function of Ω. **Section II gives us**

$$N = H^{-1} e^{-3\Omega} (12\pi R_o^3). \tag{3.5}$$

We can use the equation $dt = -N d\Omega$ and the fact that H is constant to recover the more usual solution for this metric. We find

$$t = (4\pi R_o^3/H) e^{-3\Omega}. \tag{3.6}$$

2. Quantum Behavior

With our action (2.20) and the Hamiltonian (3.1) we have a problem which can be quantized. To do this we chose the commutation relations

$$[\beta_a, p_b] = i\delta_{ab}, \tag{3.7}$$

which can be satisfied by taking

$$p_+ = -i \frac{\partial}{\partial \beta_+} \; , \; p_- = -i \frac{\partial}{\partial \beta_-} \quad . \tag{3.8}$$

The question which now arises is how to handle the Hamiltonian which can be written as a square root. There are, however, several well-known approaches to a square-root Hamiltonian. Because of the simplicity of the type I equations each of the approaches can be carried through, which is not true in more complicated cases. Therefore we shall examine **these** approaches as applied to type I universes. They are

1) The square-root approach of Bethe, Schweber, and de Hoffmann.[38]

2) The Dirac approach.

3) The Schrödinger-Klein-Gordon (SKG) approach

In the first approach we write

$$i \frac{\partial \psi}{\partial \Omega} = \pm \left(- \frac{\partial^2}{\partial \beta_+^2} - \frac{\partial^2}{\partial \beta_-^2} \right)^{1/2} \psi \tag{3.9}$$

and take the square root by means of a Fourier transform. If we write

$$\psi = \int e^{i(p'_+ \beta_+ + p'_- \beta_-)} \phi(p'_+, p'_-) \frac{dp'_+ \, dp'_-}{(2\pi)} \tag{3.10}$$

and let $\pm \left(-\frac{\partial^2}{\partial \beta_+^2} - \frac{\partial^2}{\partial \beta_-^2} \right)^{1/2}$ operate on ψ we find

$$H\psi = \pm \int \sqrt{p'^2_+ + p'^2_-} \; e^{i(p'_+ \beta_+ + p'_- d\beta_-)} \phi(p'_+, p'_-) \frac{dp'_+ dp'_-}{(2\pi)} \; . \tag{3.11}$$

For plane-wave states we take $\phi = \delta(p'_+ - p_+)\delta(p'_- - p_-)$ and we find $\psi = A \, e^{i(p_+ \beta_+ + p_- \beta_- - E\Omega)}$ with $E = \pm \sqrt{p_+^2 = p_-^2}$ (p_+, p_-, E constant).

We shall next discuss the Dirac approach. This has been considered independently by Jacobs, Misner, and Zapolsky[41]. If we linearize the square-root of the operator $\dfrac{\partial^2}{\partial\beta_+^2} + \dfrac{\partial^2}{\partial\beta_-^2}$ we find that we want

$$i\,\frac{\partial\psi}{\partial\Omega} = i\alpha_+\,\frac{\partial\psi}{\partial\beta_+} - i\alpha_-\,\frac{\partial\psi}{\partial\beta_-} \tag{3.12}$$

where α_+ and α_- are matrices satisfying $\alpha_+^2 = \alpha_-^2 = 1$, and α_+ and α_- anticommute. The minimal rank for these matrices is two, and we find that any two Pauli spin matrices will do for α_+ and α_- (we shall choose $\alpha_+ = \begin{bmatrix} 0 & 1 \\ 1 & 0 \end{bmatrix}$, $\alpha_- = \begin{bmatrix} 0 & i \\ -i & 0 \end{bmatrix}$). We find that ψ must be a two-component spinor. If we write $\psi = \begin{pmatrix} \psi_+ \\ \psi_- \end{pmatrix}$, we find plane-wave solutions with

$$\psi_\pm = A_\pm\,e^{i(p_+\beta_+ + p_-\beta_- - E\Omega)}\,, \qquad (E,\ p_+,\ p_-\ constants) \tag{3.13}$$

with $E = \pm\sqrt{p_+^2 + p_-^2}$. For each of these values of E we find a single solution for A_+ and A_-, $A_+ = \dfrac{1}{\sqrt{2}}\,A_- = -\dfrac{(p_+ - ip_-)}{\sqrt{2}\,E}$. These solutions, as is to be expected, are similar to the square-root solution, but the fact that there are the two spinor degrees of freedom is disturbing. The two solutions for E are easily interpretable in terms of expanding ($E<0$) and contracting ($E>0$) universes, but the interpretation of the "spin" states ψ_+ and ψ_- is not straightforward in terms of any known physical attributes of the universe.

The final approach we shall consider is SKG quantization, where we write from Eq. (3.1),

$$-\frac{\partial^2\psi}{\partial\Omega^2} + \frac{\partial^2\psi}{\partial\beta_+^2} + \frac{\partial^2\psi}{\partial\beta_-^2} = 0 \tag{3.14}$$

This equation has the well known solutions

$$\psi = A\, e^{i(p_+ \beta_+ + p_- \beta_- - E\Omega)}, \quad (p_+, p_-, E \text{ constant})$$

with $E = \pm \sqrt{p_+^2 + p_-^2}$. Such a formulation has many advantages in this case; the equation is well known (unlike the square-root equation), the wave functions fit the fact that the universe has no "spin" degrees of freedom, and the two types of solution $(E \gtrless 0)$ correspond to expanding and contracting universes.

The disadvantages are few, the major being the difficulty in interpreting probability density, especially the fact that the usual probability density for the SKG approach, $\rho = i(\psi^* \frac{\partial \psi}{\partial t} - \psi \frac{\partial \psi^*}{\partial t})$, can become negative. On the whole, the advantages outweigh the disadvantages and this approach has come to be widely used in quantum cosmology during its short history.

If we adopt the SKG approach for Bianchi type I cosmologies, we need only construct wave packets which move along the classical trajectories and allow E to be positive or negative depending on whether we wish to examine expanding or contracting universes. The construction of such wave packets and their interpretation is not difficult.

B. The Kantowski-Sachs Universe

1. *Classical Behavior*

Fishbone[22] has studied the application of the ADM procedure to the vacuum case of a universe considered by Kantowski and Sachs[18], and we shall present his results. The universe is defined by a line element

$$ds^2 = -dt^2 + S^2(t)dz^2 + R^2(t)(d\theta^2 + sin^2\theta d\phi^2). \tag{3.15}$$

Kantowski and Sachs give a solution for this universe in the reference cited. They also show that the volume of the universe (as defined in Section II) begins at zero at some time t_o, grows to a maximum, then returns to zero, somewhat in the fashion of the Friedmann $k = + 1$ universes. Of course, the changes in geometry during this expansion and recontraction are more complex as will be discussed later.

In order to consider the ADM formalism for this metric, we shall rewrite it as

$$ds^2 = -N^2 d\Omega^2 + \frac{R_o^2}{4\pi} e^{-2\Omega} e^{2\beta}{}_{ij} \; \tilde{\sigma}^i \; \tilde{\sigma}^j , \tag{3.16}$$

where $\tilde{\sigma}^3 = dz, \tilde{\sigma}^2 = sin\theta d\phi$ and $\tilde{\sigma}^1 = d\theta$ and β_{ij} and Ω are functions of t (or Ω) only. The β matrix is *diag* $(\beta, \beta, -2\beta)$. In this form the metric resembles that of the Bianchi-type universes. As is well known, however, the Kantowski-Sachs universe is not one of the Bianchi type models since it does not admit a simply transitive group of isometries. Instead, its homogeneity group is $R \times SO(3,R)$, as is obvious from eg. (3.15).

Because we can write our metric in the form (3.16), we see that if, as we expect, ^3R for the Ω - constant surfaces, and $A^{ij}{}_{|j}$ for any homogeneous, tensor density A^{ij}, depend only on β and Ω, then the ADM procedure would proceed as in the Bianchi-type universes. In fact, if we artificially close this universe by restricting z to lie between 0 and 4π, the ADM reduction of Section IIIC may be used without modification. We need only make the restrictions $\phi = \psi = \theta = \beta_- = 0$ and let $\beta_+ \equiv \beta$ in our β-matrix of that section. This implies that $p_\phi = p_\psi = p_\theta = p_- = 0$ and if we define $p_+ \equiv p$, we have

$$H^2 = p^2 - 24\pi^2 g \, ^3R, \tag{3.17}$$

and in action

$$I = \int [p d\beta - H d\Omega] \quad . \tag{3.18}$$

The constraints $\pi^{ij}{}_{|j}$ and the scalar curvature ^3R remain to be calculated before we can be sure this procedure is consistent. Fishbone[22] has shown that the $\pi^{ij}{}_{|j}$ constraints are satisfied identically, as in the Bianchi type I universes. As in that case we can then choose $N_i = 0$ (as is implied in our choice of (3.16) as our metric) with no qualms. Fishbone[22] has also calculated ^3R and finds it to be $2(\frac{4\pi}{R_o^2})e^{2\Omega+2\beta}$. This leads to an unconstrained Hamiltonian

$$H^2 = p^2 - 3R_o^4 e^{-2\Omega+2\beta} \quad . \tag{3.19}$$

Fishbone[22] has discussed this formulation but does not give a solution. He finds it more convenient instead to make two consecutive transformations. The first of these consists of defining a new time coordinate λ such that $d\Omega/d\lambda = H \equiv -p_\Omega$ and defining $p_\beta \equiv p$. The second

has $\beta \to \frac{\sqrt{3}}{3}(2\beta + \Omega)$, $\Omega \to \frac{2}{3}\sqrt{3}(\frac{1}{2}\beta + \Omega)$. These lead to

$$I = \int \{p_\beta d\beta + p_\Omega d\Omega - [\frac{1}{2}(p_\beta^2 - p_\Omega^2 - 3R_o^4 e^{-2\sqrt{3}\Omega})]d\lambda\}, \qquad (3.20)$$

with the constraint $H = \frac{1}{2}(p_\beta^2 - p_\Omega^2 - 3R_o^4 e^{-2\sqrt{3}\Omega}) = 0$. This constraint is

easy to satisfy because H is a constant of the motion (because it is

independent of λ). The momentum p_β is also a constant of motion.

Fishbone gives as a general solution to the equations of motion:

$$\beta = p_\beta \sqrt{3} R_o^2 \lambda - \Omega_o$$

$$\Omega = \Omega_o + \frac{1}{\sqrt{3}} \ln\left[\frac{c^2 e^{6p_\beta R_o^2 \lambda} + 1}{(c^2+1) e^{6p_\beta R_o^2 \lambda}}\right], \quad c = p_\beta e^{\sqrt{3}\Omega_o} \pm \sqrt{p_\beta^2 e^{2\sqrt{3}\Omega_o} - 1}. \qquad (3.21)$$

Figure (3.1) is a sketch of the motion of the universe point for

$\Omega_{min} = 0$, where Ω_{min} corresponds to the maximum of expansion

(turnaround). Choices of constants of motion, Ω_o and c, merely place

Ω_{min} and $\beta(\Omega_{min})$ at various points on the $\beta\Omega$-plane. The eqality of H

to zero allows the following potential condition to be given

$$p_\beta^2 - \dot{\Omega}^2 - 3 R_o^4 e^{-2\sqrt{3}\Omega} = 0, \qquad (3.22)$$

where $\dot{\Omega} \equiv d\Omega/d\lambda$. Figure (3.2) shows the effect of this condition and

points up the fact that the problem is equivalent to that of a particle

scattering off a one-dimensional exponential potential in a two-dimen-

sional space. The singularities (initial and recollapse) occur at the

two points where $\Omega = \infty$. We shall discuss the nature of these singu-

larities in another section.

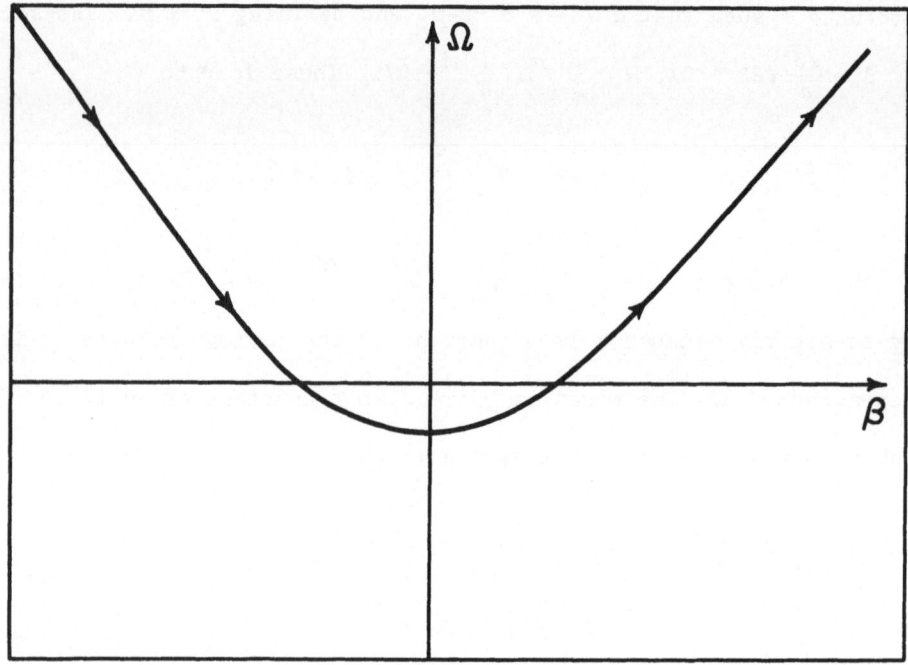

Figure 3.1. A sketch of the motion of the Kantowski-Sachs universe in the $\beta\Omega$-plane for $\Omega_{min} = 0$. (Figure courtesy L. Fishbone)

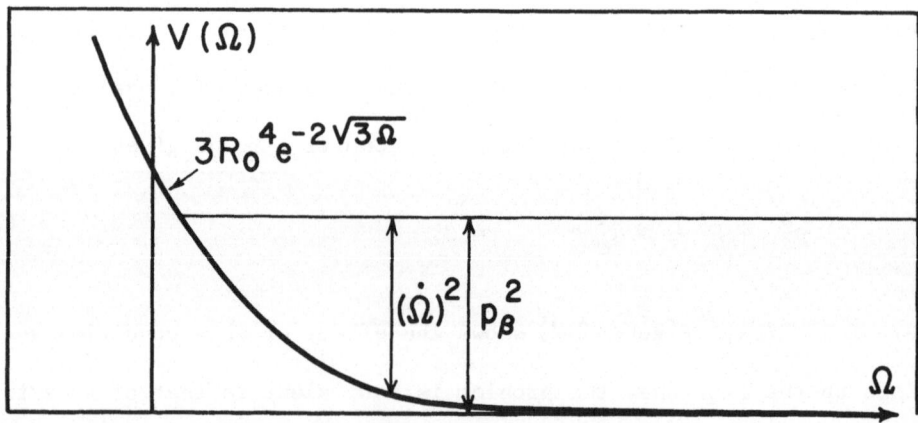

Figure 3.2. A sketch of the relation $(\dot{\Omega})^2 - (p_\beta)^2 + 3R_0^4\, e^{-2\sqrt{3}\Omega} = 0$
The horizontal line is the orbit of the universe point. There is a classical turning point where this line meets the potential $3R_O^4\, e^{-2\sqrt{3}\Omega}$. (Figure courtesy L. Fishbone)

2. *Quantum Behavior*

If we reverse one of the canonical transformations of the action
(3.20) and return to Ω-time, we see that we have, as we had in the Bianchi-
type I case, a square-root Hamiltonian. Because SKG quantization proved
so much simpler in that case, we use it here. Using the substitution
$\frac{p_\beta}{R_o{}^2} = -i\,\frac{\partial}{\partial\beta}$ we find that $R_o{}^2(-\frac{\partial^2}{\partial\beta^2} + \frac{\partial^2}{\partial\Omega^2} - 3\,e^{-2\sqrt{3}\Omega})\psi = 0$. Fishbone
has shown that this equation has solutions

$$\psi(\beta,\Omega) = e^{-i\sqrt{3}k\beta}\,F_{ik}\,(e^{-\sqrt{3}\Omega}) \quad , \tag{3.23}$$

where F_{ik} are modified Bessel functions, usually denoted by $I_\nu(z)$ or
$K_\nu(z)$ and k is a superposition parameter.

Because the classical problem resembles that of a particle scattering
from an exponential wall in two dimensions, it is best to consider the
quantum problem as a scattering problem also. Thus, if we examine the
asymptotic form of the wave function (that is far from the potential
wall), we should be able to distinguish incoming waves and outgoing,
phase-shifted waves, and be able to compute a phase shift. This has
been done by Fishbone[22], and we shall describe his computations.

The asymptotic forms of the modified Bessel functions are $I_\nu(z)$
$\rightarrow e^z/2\pi z$ and $K_\nu(z) \rightarrow \sqrt{\pi/2z}\,e^{-z}$ for large z. Because $\Omega \rightarrow -\infty$ is a
classically forbidden region we discard $I_\nu(z)$. In the asymptotic
region we find $\psi_k(\Omega,\beta) = e^{-i\sqrt{3}k\beta}K_{ik}(e^{-\sqrt{3}\Omega})$ or

$$\psi_k(\beta,\Omega) = g(k)\left[e^{-i\sqrt{3}k(\beta+\Omega)} - e^{2i\delta(k)}\,e^{i\sqrt{3}k(\Omega-\beta)}\right] . \tag{3.24}$$

where $g(k)$ is an unimportant function of k, and $e^{2i\delta(k)} = (\frac{1}{2})^{-2ik}\,\frac{\Gamma(1+ik)}{\Gamma(1-ik)}$.
The phase shifts for any k can be obtained from tabulated Γ-functions,

but in the limiting cases they are:

$$\delta(k) \xrightarrow[k \to 0]{} k(\ln 2 - \gamma) \qquad\qquad (3.25a)$$

$$\delta(k) \xrightarrow[k \to \infty]{} k\ln k, \qquad\qquad (3.25b)$$

where γ is Euler's constant.

Fishbone[22] proceeds to calculate the spreading of a wave packet that is initially Gaussian in k-space with a width Δ and centered at $k = \bar{k}$, that is $\psi_{in} = \int_{-\infty}^{\infty} f(k)\ e^{ik\sqrt{3}\ (\beta+\Omega)}\ dk$, where $f(k) = e^{-(k-\bar{k})^2\Delta^2/4}$. For such a situation the outgoing wave satisfies

$$\psi_{out} \to -\int_{-\infty}^{\infty} f(k)\ e^{2i\delta(k)}\ e^{i\sqrt{3}k(\Omega-\beta)}\, dk \quad . \qquad\qquad (3.26)$$

For the low frequency case $(\bar{k} \to 0)$ we find that the wave packet corresponding to the classical solution (3.21) comes in along a line of slope -1 through the point $\Omega = \beta = 0$ and exits along a line of slope $+1$ through the point $[\beta = 0, \Omega = -\frac{2}{\sqrt{3}}(\ln 2 - \gamma)]$ with no spreading. Figure (3.3) shows this behavior. In the high frequency case $(\bar{k} \to \infty)$, the integration of (3.26) is more difficult, but Fishbone[22] displays solutions for ψ_{in} and ψ_{out} which follow roughly the same track as in the low-frequency case. These solutions exhibit spreading unless $\Delta^4\bar{k}^2>>1$, that is at lower frequencies. Note, however, that in both the low and high frequency cases the solution is translationally invariant in the $\beta\Omega$-plane.

Quantum mechanically, then, the solution is very similar to the classical one. The major interesting feature is that the universe is more "quantum mechanical" when Ω is small (near the present) than it is when it is near the singularities. That is, the region in which wave packets spread significantly is the region where the potential is large, near turnaround. This type of behavior will be seen in other universes later.

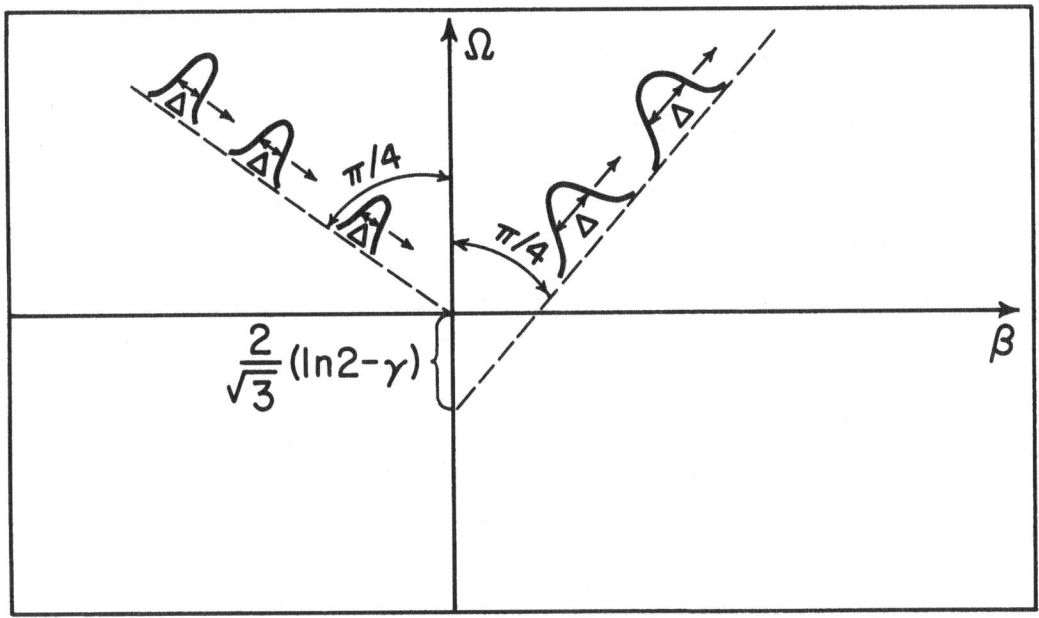

Figure 3.3. A schematic representation of the motion of a wave packet along a classical trajectory. The wave packet is represented by the bell curves which move in the direction of the arrows. The width of the packet is Δ. (Figure courtesy L. Fishbone)

C. Matter in Homogeneous Cosmologies

Up to this point we have not needed to introduce matter into the universes we have considered. As we shall show, the inclusion of fluid matter in the two cosmologies we have studied so far does not lead to any new behavior. As we study further models we shall find matter playing an increasingly important role in their behavior. Therefore, it is appropriate at this point to consider ways of including matter in homogeneous cosmologies.

We shall consider three types of material source terms for homogeneous cosmologies. These are

    1) Electromagnetic Fields

    2) Massive Vector Meson Fields

    3) Perfect Fluids.

In order to put these types of matter into the ADM formalism we need a Lagrangian for each of them . Lagrangians for electromagnetic fields and for massive vector meson fields are well known, and we can insert them without trouble. If we are able to solve the equations of motion of the fields for the field quantities in terms of the metric and constants of motion (as we can in some special cases) then these Lagrangians merely add terms to the Hamiltonian constraint C^o and the space constraints C^i. There exists at present no satisfactory Lagrangian formalism to handle pure fluid matter (although work by Schutz[28] suggests that one may soon be forthcoming), so we handle such matter by means of a special Lagrangian constructed to solve our specific problem. We shall describe this Lagrangian below.

We shall not give any discussion of massive vector meson fields, as work in this area has not proceeded far enough to give a useful description of the influence of such fields on the behavior of the universe. The work done so far is due to Hughston and Jacobs[42], who describe the equations of motion for these fields in Bianchi-type cosmologies.

In the reference just cited, Hughston and Jacobs also investigate the behavior of source-free, homogeneous, electromagnetic fields in Bianchi-type cosmologies. The cases they consider are

    1) Vanishing Poynting vector;

    2) Pure electric (magnetic) fields

For the solution with vanishing Poynting vector (i.e., $\vec{B} \parallel \vec{E}$) they find

$$B_i = sin(\xi) \ e^{2\Omega} \ e^{\beta}_{ai} \ D_a \tag{3.27}$$

$$E_i = cos(\xi) \ e^{2\Omega} \ e^{\beta}_{ai} \ D_{a},$$

where

$$\xi = -\frac{1}{2} \frac{D^a D_a}{D^2} \varepsilon_{b \ lm} \ C^j_{lm} \int N e^{\Omega} \ e^{\beta}_{ib} \ e^{\beta}_{ji} \ d\Omega, \tag{3.28}$$

where C^i_{jk} are the structure constants of the particular Bianchi type considered, and the D^a are three constants subject to the constraint

$$D^a \ C^i_{ia} = 0. \tag{3.29}$$

In the pure magnetic case they find

$$B_i = e^{2\Omega} \ e^{\beta}_{ai} \ K_{a}, \tag{3.30}$$

where the K_a are three constants subject to $K_a \ C^i_{ia} = 0$. There is an additional constraint $e^{\beta}_{ij} \ B_j \ C^i_{lm} = 0$. These are used to show that there can be no pure magnetic field in types VIII, IX, IV, V, VI ($h \neq -1$) and VII ($h \neq 0$). Types III, VI ($h = -1$) and VII ($h = 0$) allow only one independent component of the magnetic field. Type II allows two independent magnetic field components while type I puts no constraints on the B_i. These results hold true for a pure electric field.

Once we have these solutions for the fields we may insert them in the standard Lagrangian density for the electromagnetic field,

$$L_M = - \frac{1}{16} (F_{\mu\nu} F_{\alpha\beta} g^{\alpha\nu} g^{\beta\mu}), \tag{3.31}$$

and obtain the modifications of the constraints which should come about.

In order to construct a Lagrangian for fluid matter, we shall first examine the equations of motion for fluids in homogeneous cosmologies for which the metric is factorable as $ds^2 = -dt^2 + R_o^2 e^{-2\Omega} e^{2\beta}_{ij} \tilde{\sigma}^i \tilde{\sigma}^j$. If we start with the equations $T^{\mu\nu}{}_{;\nu} = 0$, Misner and Sharp[43] have shown that for a fluid stress tensor, $T^{\mu\nu} = (\rho+p) u^\mu u^\nu + p g^{\mu\nu}$, these equations (if we assume baryons are conserved) are equivalent to

$$(n u^\mu)_{;\mu} = 0 \tag{3.32a}$$

$$u_{\mu;\nu} u^\nu = -(\delta^\nu_\mu + u_\mu u^\nu) \frac{p_{,\nu}}{\rho+p} \tag{3.32b}$$

$$u^\mu s_{,\mu} = 0 \quad , \tag{3.32c}$$

where n is the baryon number density and s the specific entropy. For homogeneous cosmologies we must have $s_{,i} = 0$ so Eq. (3.32c) implies $\dot{s} = 0$, or that the solutions must be isentropic. If we examine the space part of the Euler equations, $u_{i;\nu} u^\nu = -(\delta^\nu_i + u_i u^\nu) \frac{p_{,\nu}}{\rho+p}$ and define $h = h(n) \equiv (\rho+p)/n$, we find that $(hu_i)_{;\nu} (hu^\nu) = 0$. The

equation of continuity is $(nu^\mu)_{;\mu} = 0$, so if we define a vector

$w^\mu = hu^\mu$, these two equations and the normalization condition for

u^μ become

$$w_{i;\nu} w^\nu = 0 , \tag{3.33a}$$

$$\left[\frac{n}{h} w^\mu\right]_{;\mu} = 0 , \tag{3.33b}$$

$$w^\mu w_\mu = -h^2. \tag{3.33c}$$

If we consider fluids whose equations of state take the form

$p = (\gamma-1)\rho$ (for which $n = \rho^{1/\gamma}$), notice that Equation (3.33a)

can be very difficult to solve, but that equation (3.33b) in the frame

$(d\Omega,\tilde{\sigma}^i)$ is readily soluble as

$$(N\rho^{1/\gamma} u^0 R_0^3 e^{-3\Omega})^{\cdot} = 0 . \tag{3.34}$$

This implies that

$$\rho = \mu N^{-\gamma}(u^0)^{-\gamma} R_0^{-3\gamma} e^{3\gamma\Omega} \tag{3.35}$$

where μ is a constant. We want to show that without solving Eqs. (3.33a)(3.33c)

further we can construct a Lagrangian for fluid matter which has an

equation of state $p = (\gamma-1)\rho$, if we assume that ρ and u_μ are known

functions of time only, and that L_M contains no derivatives of the metric.

The second assumption is justified on the grounds that all a Lagrangian must do is produce the desired equations when varied. If this can be achieved with a simple Lagrangian, there is no need for a more complicated one. With these derivatives absent, variations of L_M with respect to $g_{\mu\nu}$ are given by partial derivatives with respect to $g_{\mu\nu}$.

With the assumption that ρ and u_μ are functions of Ω only, we can reduce $T_{\mu\nu}$ to a function of time, N, N_i, and $g_{ij} = e^{-2\Omega}\, e^{2\beta}{}_{ij}$. Because we can do this, if we can write $\partial/\partial^4 g_{\mu\nu}$ in terms of $\partial/\partial N$, $\partial/\partial N^i$, and $\partial/\partial g^{ij}$, we can construct a set of partial differential equations for L_M depending only on N, N_i, g_{ij}, ρ, and u_μ. We can make the change of variables from $^4 g_{\mu\nu}$ to N, N_i, and g_{ij} easily, and we find, for example,

$$\frac{\partial}{\partial g^{00}}\bigg|_{g^{0i}{}_4 g^{ij}} = \frac{\partial N}{\partial g^{00}}\bigg|_{g^{0i}{}_4 g^{ij}} \frac{\partial}{\partial N} + \frac{\partial N^i}{\partial g^{00}}\bigg|_{g^{0i}{}_4 g^{ij}} \frac{\partial}{\partial N^i}$$

$$+ \frac{\partial g^{ij}}{\partial g^{00}}\bigg|_{g^{0i}{}_4 g^{ij}} \frac{\partial}{\partial g^{ij}} \qquad\qquad (3.36)$$

$$= \frac{1}{2}(N)^3 \frac{\partial}{\partial N} + (N)^2 N^i \frac{\partial}{\partial N^i} + N^i N^j \frac{\partial}{\partial g^{ij}}.$$

From Eq. (2.21) the equations that L_M must satisfy are then

$$\frac{1}{2}(N)^3 \frac{\partial L_M}{\partial N} + (N)^2 N^i \frac{\partial L_M}{\partial N^i} + N^i N^i \frac{\partial L_M}{\partial g^{ij}} = -8\pi N R_o^3\, e^{-3\Omega}[\gamma\rho(u_o)^2-(\gamma-1)\rho(N^2-N^i N_i)] \quad (3.37a)$$

$$\frac{1}{2}(N)^2 \frac{\partial L_M}{\partial N^i} + N^i \frac{\partial L_M}{\partial g^{ij}} = -8\pi N R_o^3\, e^{-3\Omega}[\gamma\rho\, u_o\, u_i + (\gamma-1)\rho N_i] \qquad (3.37b)$$

$$\frac{\partial L_M}{\partial g^{ij}} = - 8\pi NR_o^3 \ e^{-3\Omega}[\gamma\rho \ u_i \ u_j + (\gamma-1)\rho \ g_{ij}] \quad . \tag{3.37c}$$

Using (3.37c) and (3.37b) and the equations $u_\mu \ u^\mu = - 1$ and $u^\mu = g^{\mu\nu} \ u_\nu$, we can eliminate $\frac{\partial L_M}{\partial N^i}$ and $\frac{\partial L_M}{\partial g^{ij}}$ from (3.37a) and reduce it to an equation for $\frac{\partial L_M}{\partial N}$, which we can solve readily. It then becomes possible to solve (3.37b) and give a complete solution for L_M. If we use the constant μ to eliminate ρ,

$$\begin{aligned}
L_M = &- 16\pi N\mu \ R_o^{3(1-\gamma)} e^{3(\gamma-1)\Omega} [\gamma(1+R_o^{-2} \ e^{2\Omega} \ e^{-2\beta}_{\ ij} \ u_i \ u_j)^{(1-\frac{1}{2}\gamma)} \\
&- (\gamma-1)(1+R_o^{-2} \ e^{2\Omega} \ e^{-2\beta}_{\ ij} \ u_i \ u_j)^{-\gamma/2}] \\
&- 16\pi \ N_i \ R_o^{3(1-\gamma)} \ e^{3(\gamma-1)\Omega} \ \gamma\mu \ (1+R_o^{-2} \ e^{2\Omega} \ e^{-2\beta}_{\ ij})^{\frac{1}{2}(1-\gamma)} \ u_j \ g^{ij} \quad (3.38)
\end{aligned}$$

Now that we have this Lagrangian we may specialize to any type of matter with the equation of state we have discussed. Notice that this Lagrangian adds no new degrees of freedom, so it merely serves to modify the constraints in the manner discussed in Section II. In the cases where we may always take the fluid to be co-moving, that is for which the $C^i = 0$ are identically satisfied and $u^i = 0$, the Lagrangian becomes

$$L_M = -8\pi N \ R_o^{3(1-\gamma)} \ e^{3(\gamma-1)\Omega} \mu \quad . \tag{3.39}$$

Hughston and Shepley[44] have shown that the Lagrangian for many non-interacting fluids is $\sum_n L_{M_n}$, where $L_{M_n} = - 8\pi N \ R_o^{3(1-\gamma_n)} \ e^{3(\gamma_n-1)\Omega} \mu_n$.
This form of the Lagrangian shows why we have not been much concerned

with matter so far. In Bianchi type I universes and Kantowski–Sachs universes for which $C^i = 0$ is an identity, fluid matter would only modify the Hamiltonian by adding a simple function of time which in general would not have added anything new.

We must now ask how Lagrangians such as the ones for fluid matter appear in the Hamiltonian. The terms $NL^o{}_M + N_i \, L^i{}_M$ mean that $C^o \rightarrow C^o + L^o{}_M$ and $C^i \rightarrow C^i + L^i{}_M$. From the difinition of H we find that in material-filled universes

$$H^2 = H^2_{empty} - 24\pi^2 g^{1/2} \, L^o{}_M \qquad\qquad (3.40)$$

and

$$\pi^{ij}{}_{|j} = \frac{1}{2} \, L^i{}_M \qquad\qquad (3.41)$$

D. Friedmann-Robertson-Walker Universes

1. *Classical Behavior*

We shall not devote much space to the discussion of the classical behavior of the Friedmann-Robertson-Walker universes as this subject has been throughly thrashed in the nearly fifty years it has been in existence. We shall only point out how the ADM procedure gives the well known classical solution.

The Friedmann-Robertson-Walker universes are Bianchi types I, V, and IX with $\phi = \psi = \theta = \beta_+ = \beta_- = 0$. If we look at our general Hamiltonian, we see the power of the ADM method. With these restrictions all dynamical variables disappear and the Hamiltonian reduces to

$$H^2 = -24\pi^2 \, g^3 \, R - 24\pi^2 \, g^{1/2} \, L^o_M \tag{3.42}$$

and the space constraints are identically satisfied. ^3R is well known and we find

$$^3R = \frac{3}{2} \, R_o^{-2} \, k \, e^{2\Omega} \tag{3.43}$$

where $k = 0$ for type I, $k = -1$ for type V, and $k = +1$ for type IX. We can obtain expressions for L^o_M for fluids with equations of state $p = (\gamma-1)\rho$ from Section III.C by letting $u_i = 0$ (because the space constraints are identically satisfied we want $L^i_M = 0$) and $\beta_{ij} = 0$ in our expression for the general fluid Lagrangian there. This leads to

$$L^o_M = - 16\pi\mu \, R_o^{3(1-\gamma)} \, e^{3(\gamma-1)\Omega} \tag{3.44}$$

for a single-component fluid (we shall not consider multicomponent
fluids in this section). In the ADM context, then, the problem is
already solved. We have H as a function of Ω, our time variable. The
usual Friedmann-Robertson-Walker solution is achieved in our case by
integrating $\frac{d\Omega}{dt} = -\frac{1}{N}' = H\,e^{3\Omega}/12\pi R_o^3$, to find Ω as a function of cosmic
time t.

2. *Quantum Behavior*

We shall consider the ADM **formulation of this problem and compare**
it with the only other extant quantum cosmological model, that of
DeWitt[6].

While in the ADM formalism, classically, there is no dynamical
freedom in the Hamiltonian, SKG quantization leads to a differential
equation in Ω. For a one component fluid we find

$$-\frac{d^2\psi}{d\Omega^2} = 36\pi^2\,R_o^4\,k\,e^{-4\Omega}\psi + 384\pi^3\mu\,R_o^{3(\gamma-2)}\,e^{-3(2-\gamma)\Omega}\psi. \qquad (3.45)$$

When $k = 0$ or -1 the potential term falls exponentially from zero at $\Omega =$
$+\infty$ to $-\infty$ at $\Omega = -\infty$. Since this problem corresponds to the
classical one of a particle moving in this potential along the zero
line, the singularity is not a classically forbidden region, so the
wave function need not be zero there.

Nutku[27] has studied the problem of the closed universes ($k=+1$),
both empty and containing radiation, by treating this problem as a

limiting case of the Kantowski-Sachs universe with $\beta = 0$. In that case if

we separated our wave equation, we would get a separation parameter m^2 and

our equation for ψ_Ω would look like

$$- \frac{d^2 \psi_\Omega}{d\Omega^2} = - m^2 \psi_\Omega + V(\Omega) \psi_\Omega .$$ (3.46)

Nutku[27] has retained the term m^2, writing as the general equation for

closed, radiation-filled universes,

$$- \frac{d^2 \psi}{d\Omega^2} = - m^2 \psi - 36\pi^2 R_o^4 e^{-4\Omega} \psi + 384\pi^3 R_o^2 \Gamma e^{-2\Omega} \psi ,$$ (3.47)

where we have replaced μ by the more usual Γ. The physical solutions

in this prescription are obtained by letting $m \to 0$. For convenience in

this section we shall let $R_o^2 = 6\pi$ and $\Gamma \to \Gamma / 64\pi^2$ from here on.

In the empty case ($\Gamma = 0$) we see that the problem is almost identical

to that of the Kantowski-Sachs case for $\beta = 0$, so we see from Eq. (3.23)

that we want

$$\psi = \lim_{m \to 0} K_{\frac{im}{2}} (\tfrac{1}{2} e^{-2\Omega}) .$$ (3.48)

As in the Kantowski-Sachs case, we want to break this up into incoming

and outgoing plane-wave (asympotoic Hankel function) states and compute

the phase shift, $\delta(m)$, between these two waves. Nutku[27] gives

$-\delta(m) = m(\ln 2 + \frac{1}{2} \gamma) \simeq m$. He also defines a formal scattering length

$a = \lim_{m \to 0} \left[- \frac{\delta(m)}{m} \right]$, a constant in this case equal to approximately .98.

Nutku[27] compares the meaning of this length to the meaning of the usual scattering length in atomic physics.

For the radiation-filled case the potential takes the form given in Fig.(3.4). Note that it has an absolute minimum at $\Omega = -\frac{1}{2} \ln (\frac{1}{2} \Gamma)$. Nutku[27] gives as a solution to this problem which vanishes at $\Omega = -\infty$,

$$\psi = \lim_{m \to 0} e^{\Omega} W_{\Gamma/4, \frac{im}{2}} (e^{-2\Omega}). \qquad (3.49)$$

where $W_{ab}(x)$ is a Whittaker function of orders a and b of the argument x. For $m \geq 0$ we can obtain asymptotic solutions near the singularity consisting of incoming and outgoing waves. Again Nutku[27] computes a phase shift and finds that as $m \to 0$.

$$-\delta(m) = -\frac{m}{2} \Psi (\frac{1}{2} - \frac{\Gamma}{4}) , \qquad (3.50)$$

where $\Psi(z)$ is the diagamma function. The behavior of the scattering length for this problem is sketched in Fig. (3.5).

By analogy to problems of atomic physics, the fact that the scattering length blows up seems to indicate that bound states are coming into play. Nutku[27] has investigated such bound states by letting $b^2 = m^2$ $(b^2 \geq 0)$ and investigating the solutions of Eq. (3.47) as confluent hypergeometric series, that is,

$$\psi = \exp (-\frac{1}{2} e^{-2\Omega}) e^{-6\Omega} F (\frac{1}{2} - \frac{\Gamma}{4} + \frac{b}{2}, 1 + b, e^{-2\Omega}). \qquad (3.51)$$

For this series to terminate we must have $\frac{b}{2} + \frac{1}{2} - \frac{\Gamma}{4} = -n$ (n a positive integer) and the resulting functions are generalized Laguerre polynomials. In order to insure the finiteness of the wave function, we must have $b \geq 0$, which implies that

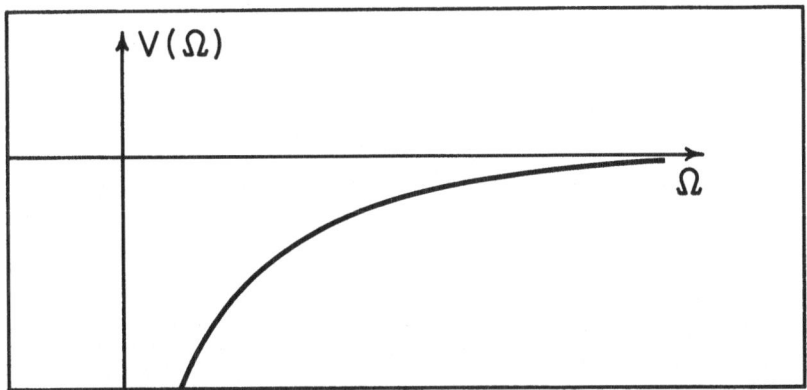

Figure 3.4a. The potential for $k = 0$ or $k = -1$ Friedmann-Robertson-Walker universes containing matter or radiation. (Figure courtesy Y. Nutku).

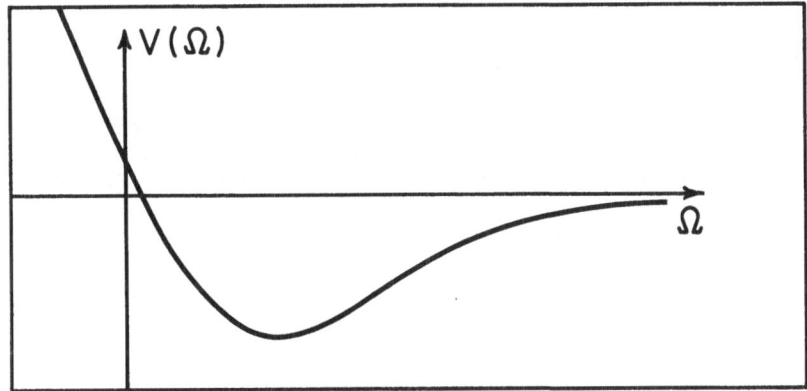

Figure 3.4b. The potential for $k = +1$ Friedmann-Robertson-Walker universes containing matter or radiation. (Figure courtesy Y. Nutku).

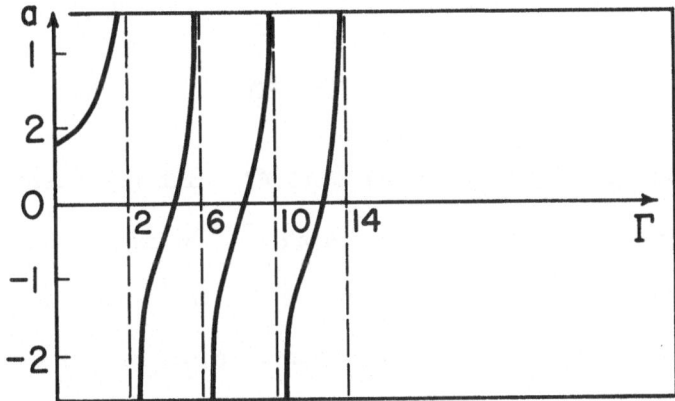

Figure 3.5. The behavior of the scattering length a for the $k = +1$ Friedmann-Robertson-Walker Universes. (Figure courtesy Y. Nutku).

$$0 < 2n \leq \frac{\Gamma}{2} - 1.$$

This means that the number of bound states is the finite number n. If we consider the limit $b \to 0$ we find that if $\Gamma = 4(n + \frac{1}{2})$, the zero-energy scattering state will also be a bound state. This is also the condition for a to blow up.

As a final note, at the values of Γ which give zero scattering length the potential behaves like a perfectly reflecting mirror.

We wish to compare the ADM formulation for the Friedmann-Robertson-Walker universes with that of DeWitt. The major differences will be seen to be in factor ordering, and in a choice of boundary conditions.

In Ref. [6] DeWitt gives the following equation for the wave functional of Friedmann-Robertson-Walker, $k = + 1$ universes,

$$\frac{1}{48\pi^2} \, R^{-1/4} \, \frac{d}{dR} \{R^{-1/2} \, [\frac{d}{dR} \, (R^{-1/4} \, \psi)]\} \, - \, 12\pi^2 \, R\psi + Nm \, \psi = 0, \quad (3.52)$$

where N is the total number of (dust) particles in the universes and m the average mass of a particle, and where $R = R_o \, e^{-\Omega}$. If we convert to derivatives in Ω-time we find that this equation reads

$$\left[\frac{d^2}{d\Omega^2} + 2 \frac{d}{d\Omega} - 60\pi^4 R_o^4 e^{-4\Omega} + 48\pi^2 NmR_o^3 Nme^{-3\Omega} + \frac{7}{16}\right]\psi = 0. \quad (3.53)$$

This is refreshingly similar to the Eq.(3.47) which was obtained by means of the ADM procedure, the term containing $e^{-4\Omega}$ being the curvature term and the one containing $e^{-3\Omega}$ the potential due to dust. The only difference in the two equations is the appearance in the DeWitt equation of both a term in $\frac{d\psi}{d\Omega}$ and a constant potential term $7/16\psi$. The constant potential $7/16$ changes ψ by very little, it corresponds to $m^2 = 7/16$ as

the physical condition rather than $m^2 = 0$ in Nutku's procedure. The

term in $\frac{d\psi}{d\Omega}$ can change the detailed form of the wave function but not in

any way which affect crucial questions such as, for example, the existence

of a singularity. The appearance of these two terms is a result of factor

ordering. If we were to rewrite the classical ADM Hamiltonian as

$R_0 \frac{3}{e^{-3\Omega}} \frac{H^2}{} + R_0 e^{-\Omega}$, and reorder the factors in the first term when we pass to

$H = -i \frac{d}{d\Omega}$ we would obtain DeWitt's equation.

The only other difference between the ADM treatment and that of

DeWitt is in choice of boundary conditions. Both Nutku and DeWitt chose

$\psi = 0$ at $\Omega = -\infty$ because of the classically forbidden region there. Be-

cause, however, DeWitt uses R for his expansion variable rather than Ω,

the point $R = 0$ represents a cutoff for R if we want R to run from $+\infty$

to $-\infty$. To exclude negative R DeWitt postulates an infinitely steep wall

at $R = 0$ which forces ψ to be zero there. In the ADM approach, if we use

Ω-time there seems no reason for postulating the existence of a hard wall

at $\Omega = +\infty$. From the form of the equation, it seems more reasonable to

let ψ be the same as that of a free-running particle. The difference

between these two choices of boundary condition hinges on "seem", because

each solution could be rewritten in Ω or R coordinates easily. The difference

in appearance of the two equations for the two coordinates is the only

indication which leads us to choose the boundary conditions we have

taken. Without some experiment to allow us to say which of these solu-

tions corresponds to the real universe, only philosophical considerations

allow us to choose between them.

E. Bianchi Type IX Universes

1. *The Diagonal Case*

a. Classical Behavior

In this case we consider type IX universes in which $d\sigma^i = \varepsilon_{ijk}\sigma^j \wedge \sigma^k$ and in which β is a diagonal metric. The procedure of Section II may be followed exactly, and, in addition, the space constraints $C^i = 0$ are again, as they were in the Bianchi type I case, identically true. Thus we may take $N_i = 0$ and obtain our Hamiltonian from the general one by letting $\psi = \phi = \theta = 0$. We have

$$H^2 = p_+^2 + p_-^2 - 24\pi^2 g\ {}^3R, \tag{3.54}$$

with our metric given by

$$ds^2 = -N^2 d\Omega^2 + R_o^2 e^{-2\Omega} e^{2\beta}{}_{ij}\ \sigma^i \sigma^j \tag{3.55}$$

Misner[21] has calculated ${}^3R$ and finds ${}^3R = 3/2\ R_o^{-2}\ e^{2\Omega}(1-V)$,

$$V(\beta) = \frac{1}{3}\ tr(e^{4\beta} - 2e^{-2\beta} + 1)$$

$$= \frac{2}{3}\ e^{4\beta+}(cosh(4\sqrt{3}\beta_-)-1) + 1 - \frac{4}{3}\ e^{-2\beta+}cosh(2\sqrt{3}\beta_-) + \frac{1}{3}\ e^{-8\beta+}\ . \tag{3.56}$$

This gives us

$$H^2 = p_+^2 + p_-^2 + 36\pi^2\ R_o^{-2}\ e^{-4\Omega}(V-1). \tag{3.57}$$

Writing $N = H^{-1}e^{-3\Omega}12\pi R_o^3$ from Section II completes our description of the metric, except that the equation $dt = -Nd\Omega$ cannot be integrated explicitly because H is not given as an analytic function of Ω.

This Hamiltonian, except for the fact that it is a square-root Hamiltonian, is equivalent to that of a particle moving in the $\beta_+\beta_-$-plane under the influence of a time-dependent potential of the form of $e^{-4\Omega}V(\beta)$.

Misner[21] has shown that $V(\beta)$ is approximately $8(\beta_+^2 + \beta_-^2)$ near $\beta = 0$

and is roughly of the shape of an equilateral triangle for large β, but

with three soft narrow channels at the corners which run out to infinity.

The equipotentials of this potential for large β were given in Figure (1.1).

Because the potential term is explicitly time-dependent, the descrip-

tion of the motion is slightly more complicated than in the usual

particle-mechanics case, but not much. The first fact we need to describe

the motion is that for large Ω, where β is small we get $H^2 \approx p_+^2 + p_-^2$

which means as it did in the Bianchi type I case, that the particle (the

universe point) moves in a straight line with velocity (in the sense of

$d\beta/d\Omega$) of unit magnitude. The second thing we need is a simplified

description of the potential. Because the walls of the triangular poten-

tial are exponentially steep, Misner[24] shows that they can be approximated by

infinitely steep walls. Because the potential is time-dependent, these

walls must move. If we consider the wall which is perpendicular to the

β_+-axis, we see that the potential in the direction of this wall is

asymptotically $V \sim \frac{1}{3} e^{-8\beta_+}$. The position of the wall can be defined by

the point where $p_+ = 0$ when the universe point is moving directly at

the wall, that is by $H \sim \frac{1}{3} e^{-4(\Omega + 2\beta_+)}$. This implies

$$\beta_+ \approx \beta_{wall} = -\frac{1}{2}\Omega - \frac{1}{8} \ln (3H^2). \qquad (3.58)$$

During those eras when the universe point is far from the walls, $H \approx const.$

and we find $d\beta_{wall}/d\Omega = -\frac{1}{2}$. Because of the triangular symmetry of the

well, all the walls have a similar motion, that is, the triangle expands

with velocity one-half. Because the velocity of the universe point is greater than that of the walls, the universe point must collide with the walls. When the universe point collides with the walls, H is no longer constant, so the wall motion is different during these collisions.

The gross features of the motion of the universe point can be given, then, by the motion of a particle bouncing around in a triangular box with straight-line motion between bounces. Note that to complete such an approximate (or *qualitative*) solution to the problem we need to know how H changes during collisions with the walls and what the law of reflection is for bounces off the walls. We shall discuss this in Section V.

While the details of the motion are not given by this approach, they can be obtained by means of a study of Hamilton's equations for particular cases. An excellent example of this is the study by Misner[24] of "corner-run solutions", in which the universe point attempts to move into one of the channels, where he shows that it cannot move far into the channel.

In the channel which surrounds the β-axis the potential $V(\beta)$ becomes asymptotically $({}^4\beta_-{}^2 e^{4\beta_+} + 1)$, so if β_- were exactly zero the particle could go out to $\beta_+ = \infty$. Misner has shown that such a situation is un- stable. He does this by assuming $\beta_+ = \beta_o + \Omega$ and taking perturbation in β_-. The Hamiltonian for β_- motion becomes

$$H^2 = p_-{}^2 + 204\pi^2 R_o{}^{-2} \beta_-{}^2 e^{4\beta_o} . \tag{3.59}$$

which is the Hamiltonian for simple harmonic motion with $\omega_- = \sqrt{204}\pi R_o{}^{-1} e^{2\beta_o}$. If we now allow β_o to change slowly, we find an adiabatic invariant

$\Sigma = H/w_- = \dfrac{H^2 \dot{\beta}_- + \frac{1}{4} w_-^2 \, \beta_-^2}{w_-}$ which combined with Hamilton's equations

for β_0 gives $\beta_0 = 2 ln(\Omega_0 - \Omega) + const$. This shows that as Ω increases

toward Ω_0 the universe point drifts away from the corner to begin

bouncing on the flat walls as soon as the approximation of small β_-

breaks down.

As a final note to the classical problem, we can use the matter

Lagrangian of Subsection III C to examine how we may put fluid matter

into diagonal type IX universes. The Euler equations for both dust

and fluid radiation in the type IX case can be seen to give $w^i = const$.

Since the space constraints are identically satisfied the form of L_M^i

implies that $u_i = 0$. This can be readily inserted in our matter

Lagrangian to give $L_M^0 = -16\pi\mu$ for dust and $L_M^0 = -16\pi \, \Gamma e^{\Omega}$ for radiation

with $L_M^i = 0$ for both. These terms add simple β-independent terms to H^2

and, therefore, are relatively uninteresting when we are discussing

motion in the $\beta_+ \, \beta_-$ -plane.

      b.   Quantum Behavior

As we begin our study of the quantum mechanical models

of type IX universes, we are in a position to make one more remark about

various methods of handling the square-root Hamiltonian. If we take our

usual commutation relations, we get

$$H^2 = -\frac{\partial^2}{\partial \beta_+^2} - \frac{\partial^2}{\partial \beta_-^2} + 36\pi^2 \, R_0^{-2} \, e^{-4\Omega} (V-1) \, . \tag{3.60}$$

We see that the term $36\pi^2 \, R_0^{-2} \, e^{4\Omega} (V-1)$ appears under the square root

when we compute H. As has been pointed out by several people[45] this means that this term is *not* a potential but actually the equivalent of a time-and-space-dependent mass in the relativistic particle problem. Therefore, if we attempt to linearize the square-root Hamiltonian by the Dirac method, when we attempt to recover H^2 by squaring our Hamiltonian linear in $\frac{\partial}{\partial\beta_+}$ and $\frac{\partial}{\partial\beta_-}$ we obtain terms involving the derivatives of V which are extremely difficult to remove by purely algebraic combinations. This again points to the SKG method as the best method for quantization.

In this case, for the first time, we have encountered both problems that we have said plague quantum cosmology, the fact that we have a squared Hamiltonian and the fact that H is explicitly time-dependent. The first problem we have eliminated by restricting ourselves to SKG quantization, the second is more difficult, but only because of the technical difficulty we have in solving the resulting differential equation. After Misner[21] and Zapolsky[13] we shall not try to solve the full problem, but attempt to investiage the simpler problem in which we replace V by its approximation in terms of a triangular potential with infinitely hard walls which expands with velocity one-half.

The problem of a triangular static well has been solved and is given in a book by Schelkunoff[46], but Misner[47] has been unsuccessful in attempting to build a solution for the time-dependent case out of these solutions. Thus, at this point, we have no general solution for even the most simplified version of the quantum theory. We can, however, make some general remarks on the nature of the solution.

Zapolsky[13] has considered the solution for a one-dimensional, expanding box. We give this solution in Appendix B , and only give here the fact that the energy eigenvalues are

$$E_n \sim n \ t^{-1} \tag{3.61}$$

where n is an integer which ranges from 1 to ∞, and t is the time. For a two-dimensional box we would expect something similar, with $E_n \sim \sqrt{n^2+m^2} \ t^{-1}$. Thus we shall take

$$E_n \sim |n| \ \Omega^{-1} \ , \tag{3.62}$$

where $|n|$ corresponds to $\sqrt{n^2+m^2}$ as the energy eigenvalues for the quantized type IX universe. We shall need this estimate in a later section, but for the present we shall not consider these energy eigenstates further.

Of somewhat more interest is the solution for a wave packet bouncing around inside the triangular well. Zapolsky[13] has studied this. In his work he considers the wave packet

$$\psi_{in} = \int f(|p|) exp \ -i(p_+\beta_+ + p_-\beta_- + \sqrt{p_+^2+p_-^2} \ \Omega) \ d^2p \ , \tag{3.63}$$

where $f(|p|)$ is a function, which is sharply peaked around some central $|p|_c$. The expansion is in terms of the eigenstates of the free(read type I) Hamiltonian. These are plane-wave (one dimensional) states of infinite extent transverse to the vector $\vec{p}$ in the $\beta_+ \ \beta_-$ -plane and of finite width in the direction of propagation. As is pointed out by Zapolsky[13] such wave packets do not spread. If we allow this packet to bounce off the wall which is perpendicular to the β_+ -axis, we can use the classical bounce law (discussed in Section V and Appendix C) to our advantage. The quantities $K = \frac{1}{2} p_+ + \sqrt{p_+^2 + p_-^2} = const$, $p_- = const$. are conserved quantum-mechanically, so the classical bounce laws hold in this case also. This means that long after the bounce

$$\psi_{out} = \int f(|p'|)exp[ip'_+ d\beta_+ + p'_- d\beta_- + \sqrt{p_+'^2 + p_-'^2} \; \Omega]d^2p. \qquad (3.64)$$

We assume that the initial state had a dimensionless width $\Delta \equiv p_+\delta\beta_+ + p_-\delta\beta_-$, where $\delta\beta_\pm$ gives the width in $\beta_+\beta_-$ -space. Because the outgoing packet has the same functional form in the variables $p'_\pm$ as the original did in $p_\pm$, the width Δ must be invariant, so

$$p_+\delta\beta_+ + p_-\delta\beta_- = p'_+ \delta\beta'_+ + p'_-\delta\beta'_- \; , \qquad (3.65)$$

or

$$\frac{\delta\beta_+^2 + \delta\beta_-^2}{\delta\beta_+'^2 + \delta\beta_-'^2} = \frac{p_+'^2 + p_-'^2}{p_+^2 + p_-^2} = \frac{H'}{H} < 1 \; . \qquad (3.66)$$

This implies that the packet spreads in $\beta_+ \beta_-$ -space. Thus, the wave packet follows the trajectory of a classical particle in the well, bouncing around, while spreading discontinuously at each bounce. If we use the fact that the area of the well is proportional to Ω^2 and the fact that $H\Omega \sim const.$ (see Appendix C), we find that the ratio of "area" of the wave packet($\sim(\delta\beta_+^2 + \delta\beta_-^2)$) to the area of the triangle is approximately constant. This means that the wave packet occupies a fixed fraction of the well for all time. In this sense the behavior of the packet does not become more "quantum-mechanical" as $\Omega \to \infty$. This point will be useful later.

2. *The Symmetric Case, Classical Behavior*

The symmetric case, where we have $\psi = \theta = 0$, $\phi \neq 0$, in the general β matrix of Section II provides the first opportunity to observe the action of the constraints $C^i = 0$. In previous examples this constraint has been identically satisfied, in this one the attempt to let $C^i = 0$ will, as we shall show, force p_ϕ to be zero and imply that ϕ is constant or, by proper initial choice, $\phi = 0$, which would reduce us to the diagonal case. This means we must introduce some sort of matter in order to satisfy the constraints by having $C'^i \equiv \pi^i{}_{|j} - \frac{1}{2} L^i{}_M = 0$. The simplest possible matter we can introduce would be dust.

The Lagrangian for dust is given in Section III.C and requires only the calculation of u^i to be used. The space part of Eq. (3.32b) for u^μ reads for the type IX case in the frame $(d\Omega, \delta^i)$,

$$u^i = \frac{1}{2} \frac{N}{(1+g^{ij} u_i u_j)} [u_o u_k N^j \epsilon_{ijk} + u_k u_\ell g^{\ell j} \epsilon_{ijk}] \qquad (3.67)$$

and this reduces to $u^i = 0$ if we choose $u_1 = u_2 = N_1 = N_2 = 0$. We shall show later that this is the proper choice of u_1, u_2, N_1, and N_2. This choice implies $u_3 = \bar{C}$, where $\bar{C}$ is some constant. With this choice we find from Sec. III C, $L^1{}_M = L^2{}_M = 0, L^3{}_M = 16\pi\mu \bar{C} R_o^{-2} e^{2\Omega} e^{4\beta_+}$. Ryan[12] has shown that for all type IX universes the quantity $\pi^{ij}{}_{|j}$ is given by

$$\pi^{ij}{}_{|j} = -\frac{1}{4} e^{-2\beta}{}_{im} [\pi^{**}, e^{2\beta}]_{jk} \epsilon_{mjk} , \qquad (3.68)$$

where π^{**} is the matrix π^{ij}, and $[A,B]$ is the commutator of matrices A and B. Inserting the symmetric-case expression for β, we find that

$$\pi^{1j}\big|_j = \pi^{2j}\big|_j = 0 \tag{3.69a}$$

$$\pi^{3j}\big|_j = -\frac{1}{4\pi} , R_o^{-2} e^{2\Omega} e^{4\beta} p_\phi . \tag{3.69b}$$

This implies that $C'^3 = \frac{3}{2} e^{2\Omega} e^{4\beta+}[p_\phi - 32\pi^2\mu \bar{C}]$, and $C'^i = 0$ implies $p_\phi = 32\pi^2 \mu \bar{C}$. This show that our choices for the u^i were consistent if we can show that our Hamiltonian is cyclic in ϕ.

Before we begin to compute the Hamiltonian we point out that the symmetric type IX universe is the first we have encountered with non-zero rotation. The rotation tensor of Ehlers[56] is zero automatically for Bianchi type I, and Friedmann-Robertson-Walker universes, and the Kantowski-Sachs universe. For type IX universes the rotation tensor is given by Ryan[12]. The symmetry of the diagonal case gives $\Omega_{\mu\nu} = 0$, but the symmetric case has $\Omega_{\mu\nu} \neq 0$, and $\Omega_{\mu\nu}$ is related to $\bar{C}$.

In order to compute the Hamiltonian we need L_M^O , which we can get from Section III.C. Inserting our β_{ij} and u_i into the Lagrangian we find

$$L_M^O = -16\pi\mu(1 + R_o^{-2} \bar{C}^2 e^{2\Omega} e^{4\beta+})^{1/2} . \tag{3.70}$$

We have that

$$H^2 = p_{ij} p_{ij} - 24\pi^2 g^3 R - 24\pi^2 g^{1/2} L_M^O . \tag{3.71}$$

The quantity ^3R for any type IX universe can be shown to be the same as in the diagonal case because $V(\beta)$ is invariant under similarity transformations on β_{ij}. This fact means that ^3R is the *same* for any type IX universe. It is *independent* of ϕ, ψ, θ. Now, if we insert p_{ij} from Eq. (2.16) with $\psi = \theta = p_\theta = p_\psi = 0$, we find

$$H^2 = p_+^2 + p_-^2 + \frac{3(p_\phi)^2}{\sinh^2(2\sqrt{3}\beta_-)} + 36\pi^2 R_0^4 e^{-4\Omega}(V-1) + 384\pi^3 R_0^3 e^{-3\Omega}\bar{\mu}(1+\bar{C}^2\bar{R}_0^{-2}e^{2\Omega}e^{4\beta_+})^{\frac{1}{2}} \quad (3.7?$$

Note that this Hamiltonian is cyclic in ϕ.

Before we give any exposition of the meaning of this Hamiltonian, we want to discuss the influence of the constraints on this Hamiltonian. Because the Hamiltonian is cyclic in ϕ, $\dot{p}_\phi = 0$, so the constraint $p_\phi = 32\pi^2\bar{\mu}\bar{C}$ is easily satisfied, and the constant $32\pi^2\bar{\mu}\bar{C}$ can be inserted for p_ϕ. Instead of following this straightforward procedure, however, we wish to treat this case as the first example in which the constraints are not identically satisfied and examine the three approaches to varying the total action. In this case we have

$$I = 2\pi \int p_+ d\beta_+ + p_- d\beta_- + p_\phi d\phi - H(\beta_\pm, p_\pm, p_\phi)d\Omega - N_3 C'^3 d\Omega \quad (3.73)$$

Table III.1 is a replica of Table II.1 for this special case. Here the constraints are given in specific form so we can see all the possibilities in full detail. In particular the constrained system is given in full and we can see what was meant by the vague prescriptions of Section II in that case. We see here the possibility of choosing N_3 in the constrained system to make $\phi = 0$ as was discussed in Section II.

We would like to be able to associate walls with the two new terms that have appeared in the Hamiltonian. Because the potential proportional to p_ϕ^2 is the analogue of the centrifugal potential in the Kepler problem

TABLE III.1

$$I = \int (p_\pm,\ \beta_\pm,\ p_\phi,\ \phi,\ N_3)$$

Case I

Complete Variational System

$$\delta p_+:\ \dot{p}_+ = -\frac{\partial H}{\partial \beta_+} + N_3\left(6e^{2\Omega}\,e^{4\beta_+}[p_\phi - 32\pi^2\mu C]\right)$$

$$\delta p_+:\ \dot{\beta}_+ = \frac{\partial H}{\partial p_+}$$

$$\delta \beta_-:\ \dot{p}_- = -\frac{\partial H}{\partial \beta_-}$$

$$\delta p_-:\ \dot{\beta}_- = \frac{\partial H}{\partial p_-}$$

$$\delta N_3:\ p_\phi = 32\pi^2\mu C$$

$$\delta \phi:\ \dot{p}_\phi = 0$$

$$\delta p_\phi:\ \dot{\phi} + \frac{3}{2}e^{2\Omega}\,e^{4\beta_+} + N_3 = \frac{\partial H}{\partial p_\phi}$$

Case III $p_\phi = 32\pi^2\mu C$

Constrained System

$$\dot{p}_\pm = -\frac{\partial H}{\partial \beta_+}\bigg|_{p_\phi = 32\pi^2\mu C}$$

$$\dot{\beta}_\pm = \frac{\partial H}{\partial p_\pm}\bigg|_{p_\phi = 32\pi^2\mu C}$$

Ancillary equations:

$$\dot{\phi} = -\frac{3}{2}e^{2\Omega}\,e^{4\beta_+} + N_3 + \frac{\partial H}{\partial p_\phi}\bigg|_{p_\phi = 32\pi^2\mu C}$$

$$N_3 = N_3(\Omega)$$

Case II $N_3 = 0$

All-Hamiltonian System

$$\dot{p}_\pm = -\frac{\partial H}{\partial \beta_\pm}$$

$$\dot{\beta}_\pm = \frac{\partial H}{\partial p_\pm}$$

$$\dot{\phi} = \frac{\partial H}{\partial p_\phi}$$

$$\dot{p}_\phi = -\frac{\partial H}{\partial \phi} = 0$$

Ancillary equations:

$$p_\phi = 32\pi^2\mu C$$

of Newtonian mechanics we shall call it the *centrifugal* potential (V_c).
The potential associated with the $L^O{}_M$ term we shall call the *rotation*.
potential (V_c). We shall call their respective walls the centrifugal and
rotation walls. If we make the same definition of wall as in the diagonal
case we find, that for large Ω we find $V_r \simeq 384\pi^3 R_o^2 \mu\bar{C}\, e^{-2\Omega}e^{2\beta+}$

$$\beta^r_{wall} \equiv (\beta_+)^r_{wall} = \Omega + \frac{1}{2}\, ln(H^2/384\pi^3 R_o^2 \mu\bar{C}) \qquad (3.74)$$

where β^r_{wall} is the coordinate of the rotation wall. Finally we have

$$\beta^c_{wall} \equiv (\beta_-)^c_{wall} = \frac{1}{2\sqrt{3}}\, sinh^{-1}\, (\sqrt{96}\pi\mu\bar{C}\, \mu\, \bar{C}/H) \qquad (3.75)$$

where β^c_{wall} is the coordinate of the centrifugal wall. A diagram of
these walls is given in Fig. (3.6).

3. *The General Case, Classical Behavior*

The most general type IX universe in which ψ, θ, ϕ are not zero
is complicated enough that we cannot solve the constraints fully to give
a complete Hamiltonian after the manner of ADM. We can, however, construct
the Hamiltonian and give the constraints which need to be solved, and give
some discussion of the motion of the universe for a dust-filled cosmology.

We begin by giving p_{ij} from Section II for this case. Since we are
considering the most general possible β we find

$$6p_{ij} = R^{-1}\left\{ \alpha_1\, p_+ + \alpha_2 p_- - \alpha_3\, \frac{3p_\phi}{sinh(2\sqrt{3}\beta_-}\right.$$

$$- \alpha_5\, \frac{3(p_\psi sin\phi - p_\phi cos\theta sin\phi + p_\theta cos\phi sin\theta)}{sin\theta sinh(3\,\beta_+ + \sqrt{3}\,\beta_-)} \qquad (3.76)$$

$$\left. - \alpha_4\, \frac{3(p_\theta sin^2\phi sin\theta - p_\psi sin\phi cos\phi + p_\phi cos\phi sin\phi cos\theta)}{sin\phi sin\theta sinh(3\,\beta_+ - \sqrt{3}\,\beta_-)} \right\} R$$

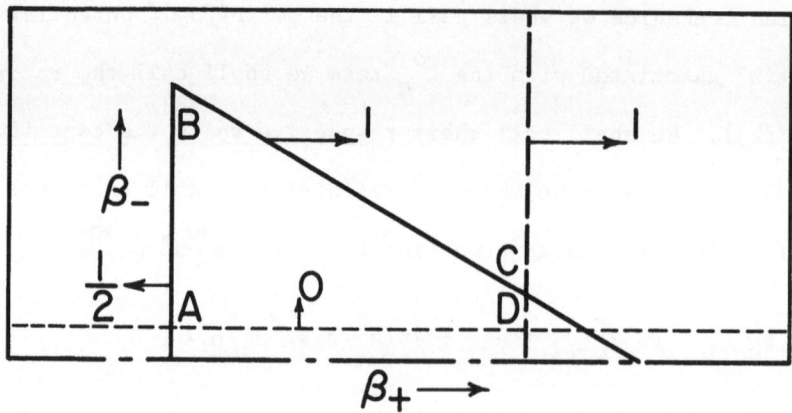

Figure 3.6.   The walls associated with the potential $V(\beta_+, \beta_-)$ (solid lines), the rotation potential (vertical dashed line), and the centrifugal potential (horizontal dotted line). The arrows give the directions and the numbers the velocities of each of the walls.   (Figure courtesy C. Misner).

With this we are able to write our Hamiltonian

$$H^2 = (2\pi)^2 (\pi^k_{\ k})^2 = 6tr(p^2) - 24\pi^2 g\ ^3R - 24\pi^2 (g)^{1/2}\ L^o_{\ M}. \qquad (3.77)$$

We can write, as in the symmetric case, ^3R as $6R_o^{-2}\ e^{2\Omega}\ (1-V)$

where, since $V = \frac{1}{3}\ tr(e^{4\beta} - 2e^{-2\beta} + 1)$ is invariant under rotations,

this is the *same* expression as in the symmetric and diagonal cases.

The three space constraints $\pi^{ij}_{\ |j} = \frac{1}{2}\ L^i_{\ M}$ are still to be satisfied.

Inserting p and e^β into Eq. (3.68), we find that the constraints to be

satisfied are

$$\{R^{-1}(\theta,\phi,\psi)p\}_i = 2\pi L_{iM} \qquad (3.78)$$

where L_{iM} is $R_o^{2\Omega} e^{-2\beta}\ e^{2\beta}_{\ ij}\ L^j_{\ M}$,

$$p = \{\frac{(p_\theta sin\phi sin\theta - p_\psi cos\phi + p_\phi cos\phi cos\theta)}{sin\theta}$$

$$\frac{(p_\psi sin\phi - p_\phi cos\theta sin\phi + p_\theta cos\phi sin\theta)}{sin\theta}, p_\phi\} . \qquad (3.79)$$

Up to this point the reduction of the general case to canonical form

has been an exercise in algebra. If we could insert L_M from Section IIIC

we could write H in closed form. Unfortunately, while we have L_M from

that section by letting $\gamma = 1$ it contains the functions u_i which were

assumed in the variational principle of that Section to be known functions

of Ω. The equation which would determine the u_i as functions of Ω is

Eq. (3.67) which is extremely complicated when β is general. This can be

seen by defining a vector $\vec{u}$ in a Euclidean space which has as its com-

ponents the components of the form $\tilde{u}$ in the σ-frame. If we choose the coordinate condition $N_i = 0$, this vector obeys the equation

$$\dot{\vec{u}} = \frac{1}{2} NR_o^{-2} e^{2\Omega} [\vec{u} \times (e^{\overrightarrow{-2\beta}} u)] / (1 + R_o^{-2} e^{2\Omega} \vec{u} \cdot e^{\overrightarrow{-2\beta}} u)^{1/2} \qquad (3.80)$$

where $\times$ is the usual cross product. If β were a constant matrix, this would be similar to the relativistic equation of motion of a rigid body with principle moments of inertia $(e^{-2\beta_+ -2\sqrt{3}\beta_-}, e^{-2\beta_+ +2\sqrt{3}\beta_-}, e^{4\beta_+})$. In our case we have a problem analogous to the problem of a solid body whose moments of inertia change with time in a complicated way. The solution of such an equation for $\vec{u}$ in terms of β and constants is, to say the least, difficult, and we shall not attempt it. We can, however, point out one constant of the motion, the analogue of $\dot{\bar{C}}$ in the previous subsection. From the form of (3.80) the magnitude of $\vec{u}$ is a constant, which we shall call $\bar{C}$ in this case also.

Without a solution of Eq(3.80) we cannot rid ourselves of u_i as a function of Ω. The existence of these unknown functions puts the problem in a form which is not quite canonical. We are, however, able to use the space constraints to produce a Hamiltonian with functions of Ω whose behavior is given by an auxiliary equation. With this we are able to give a fairly complete description of the motion of the universe.

Discussion of the Motion.

If we look at the Hamiltonian (3.77), we find that there are many points of similarity with the symmetric case. The "kinetic energy" term is

$$p_+^2 + p_-^2 + \frac{3p_\phi^2}{sinh^2(2\sqrt{3}\beta_-)} +$$

$$\frac{3(p_\psi sin\phi - p_\phi cos\theta sin\phi + p_\theta cos\phi sin\theta)^2}{sin^2\theta sinh^2(3\beta_+ + \sqrt{3}\beta_-)}$$

$$\frac{3(p_\theta sin\phi sin\theta - p_\psi cos\phi + p_\phi cos\phi cos\theta)^2}{sin^2\theta sinh^2(3\beta_+ - \sqrt{3}\beta_-)} \quad .$$

$$(3.81)$$

We see that if we could specify the behavior of θ, ϕ, and ψ, the three angle terms would have the character of centrifugal potentials as in the symmetric case and the two new ones would represent similar barriers which keep the universe point from touching axes directed along the channels in $V(\beta)$ which are at 60° to the β_+-axis. As has been pointed out, the triangular potential is the same in this case as in the symmetric and diagonal cases.

Let us now examine the potential due to matter. From Section IIIC we have for $\gamma = 1$

$$L_M = -16\pi N \mu (1 + R_o^{-2} e^{2\Omega} e^{-2\beta}{}_{ij} u_i u_j)^{1/2}$$

$$-16\pi N_i (\mu u_j g^{ij}) \quad .$$

$$(3.82)$$

We may use the constant $\bar{C}$ of the previous subsection to reduce this Lagrangian partly. We know that $u_i = \bar{C} n_i$, where n_i is a unit vector (in the sense of $n_i n_i = 1$) and $\bar{C}$ is constant. Thus L^o_M is

$$-16\pi \mu (1 + (\bar{C})^2 R_o^{-2} e^{2\Omega} n_i e^{-2\beta}{}_{ij} n_j)^{1/2} \quad .$$

$$(3.83)$$

Note that $R_{ij}(\theta, \phi, \psi) n_j$ is a unit vector also, so we may parameterize it by means of two angles γ and λ, that is

$$\vec{Rn} = (sin\gamma sin\lambda, \ sin\gamma cos\lambda, \ cos\gamma) \quad .$$

$$(3.84)$$

Noting that $\vec{n}^T e^{-2\beta} \vec{n} = (\overrightarrow{Rn})^T e^{-2\beta} d \overrightarrow{Rn}$,

$$n_i \, e^{-2\beta}{}_{ij} \, n_j = (sin^2\gamma sin^2\lambda \, e^{-2(\beta_+ + \sqrt{3}\beta_-)} + sin^2\gamma cos^2\lambda e^{-2(\beta_+ - \sqrt{3}\beta_-)}$$

$$cos^2\gamma e^{4\beta_+}) \, . \tag{3.85}$$

Hence, we have

$$L_M^0 = -16\pi \, \mu(1 + (C)^2 R_0^{-2} \, e^{2\Omega} [sin^2\gamma sin^2\lambda \, e^{-2(\beta_+ + \sqrt{3}\beta_-)}$$

$$+ \, sin^2\gamma cos^2\lambda \, e^{-2(\beta_+ - \sqrt{3}\beta_-)} \tag{3.86}$$

$$+ \, cos^2\gamma \, e^{4\beta_+}])^{1/2} \, .$$

If we examine this matter potential, we see that it defines three exponential walls in the β_+ β_--plane which close off the three channels at the
corners of the triangular gravitational potential (see Fig. (3.7)). For
large Ω these three walls correspond to the exponential potentials;
$V_r^1 \sim \bar{\mu C} R_0^3 \, e^{-2\Omega} \, e^{-(\beta_+ + \sqrt{3}\beta_-)} sin\gamma sin\lambda$, $V_r^2 \sim \mu\bar{C} R_0^3 \, e^{-2\Omega} \, e^{(\beta_+ - \sqrt{3}\beta_-)} sin\gamma cos\lambda$,
and $V_r^3 \sim \mu\bar{C} R_0^3 \, e^{-2\Omega} \, e^{2\beta_+} \, cos\gamma$. It is easy to see, if we neglect the
change of γ and λ with Ω, that these walls have the same type of motion
as the single rotation wall in the symmetric case; that is, they move with
the velocities of their respective corners and maintain a constant
distance from them. We can best interpret γ and λ if we relax the requirement that the walls be regarded as vertical. We shall represent each wall
by a plane, but a plane which makes an angle ξ_i with the normal to the
β_+ β_--plane for each V_r^i (see Fig. (3.8)). Thus, if the universe point
moves with $H = constant$ directly into the channel (we are neglecting the
fact that H does not remain strictly constant when the V_r^i are not strictly

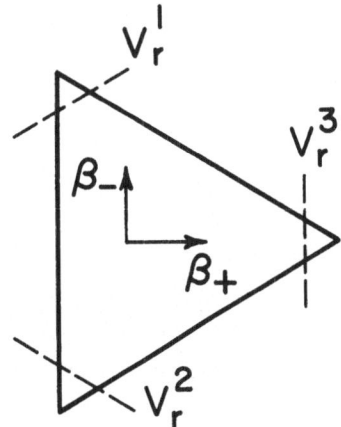

Figure 3.7. The position of the three rotation walls in the general
type IX case.

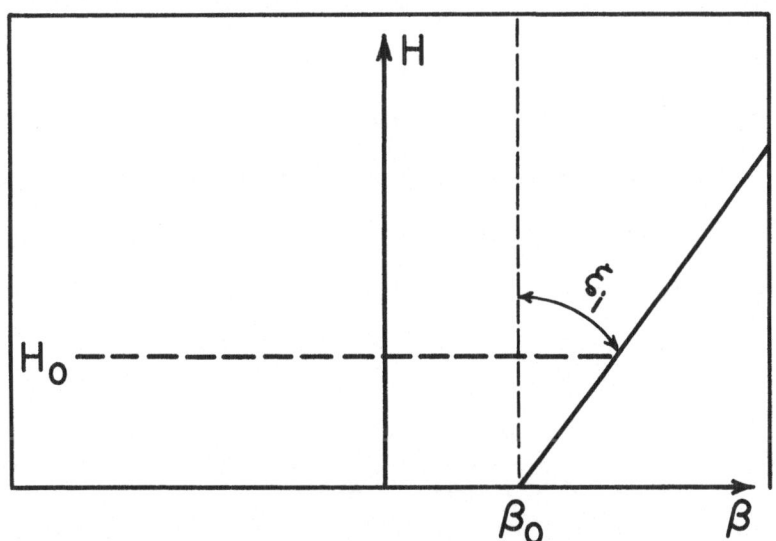

Figure 3.8. The definition of the quantities ξ_i and β_o.

zero) along the β-axis, for instance; it collides with the wall at

$\beta_w^o + H tan \xi_3$, where β_w^o is the value of β_+ at which the universe point

would have hit the wall if ξ_3 were zero. If we calculate the collision

point in the same manner as we did for the symmetric case, we find the

$\xi_3 = tan^{-1} (\frac{-ln(cos \ \gamma)}{H})$. Similarly, $\xi_1 = tan^{-1} (\frac{-ln(sin\gamma sin\lambda)}{H})$, and

$\xi_2 = tan^{-1} (\frac{-ln(sin\gamma cos\lambda)}{H})$. If one looks at these expressions, he can

see that the arguments of each function are such that for each value of

γ and λ there are three unique ξ_i.

So, in the general case, the single, hard rotation wall of the

symmetric case becomes a set of "shutters" which close off the three

channels in the $\beta_+ \ \beta_-$-plane. These shutters open and close in a manner

which depends on the dynamics of γ and λ. We can note that the three

shutters are not all independent, because their positions depend only

on two angles. For instance, if the V_r^3 shutter opens to expose the channel

along the β_+-axis, it drags a combination of the V_r^2 and V_r^1 shutters

(the exact combination is determined by the magnitude of λ) up to cover

the other two channels. In order to determine the dynamics of γ and λ,

we must look at Eq.(3.80) which ties together ψ, θ, φ, λ, and γ. We

may rewrite this equation as an equation for n_i and find

$$(\overrightarrow{Rn})^{\cdot} = (\frac{1}{6\pi})^{1/2} \frac{e^{-\Omega} R \overline{C}}{H(1+R_o^{-2}e^{2\Omega}(\overline{C})^2 (\overrightarrow{Rn})^T e^{-2\beta} d\overrightarrow{Rn})^{1/2}} \ \overrightarrow{Rn} \times (e^{2\beta} d\overrightarrow{Rn}) - R(R^{-1})^{\cdot} \overrightarrow{Rn}.$$

$$(3.87)$$

Notice that the only place where φ, ψ, and θ appear explicitly is in

the term $R(R^{-1})^{\cdot}$. We can use the space constraints and Hamilton's

equations for $\dot{\phi}$, $\dot{\psi}$ and $\dot{\theta}$ to eliminate this term. Hamilton's equations

for $\dot{\psi}$, $\dot{\theta}$, and $\dot{\phi}$ give us these quantities in terms of p_i and functions of

H, Ω, β_+ and β_-. From Eq.(3.78) and the space part of the matter Lagrangian (3.82) we have

$$p_i = 32\pi^2 \; \mu\bar{C}(Rn)_i \tag{3.88}$$

Substituting for p_i, we can now eliminate $\dot{\phi}$, $\dot{\psi}$ and $\dot{\theta}$ from the matrix $R(R^{-1})^{\cdot}$ in terms of γ and λ. We find that this procedure also eliminates *all* the angles θ, ϕ, ψ themselves; in fact we find

$$R(R^{-1})^{\cdot} = - \frac{3[32\pi^2 \; \bar{C}(Rn)_3]\kappa_3}{H\sinh^2(2\sqrt{3}\beta_-)} - \frac{3[32\pi^2 \; \bar{C}(Rn)_2]\kappa_1}{H\sinh^2(3\beta_+ + \sqrt{3}\beta_-)} - \frac{3[32\pi^2 \; \bar{C}(Rn)_1]\kappa_2}{H\sinh^2(3\beta_+ - \sqrt{3}\beta_-)} \; , \tag{3.89}$$

where $\kappa_2 = \begin{bmatrix} 0 & 0 & 1 \\ 0 & 0 & 0 \\ -1 & 0 & 0 \end{bmatrix}$.

We would like in the rest of this section to simplify the Hamiltonian by defining

$$36\pi^2 \; R_o^4 = 1,$$

$$32\pi^2 \; \mu\bar{C} = \mu C, \tag{3.90}$$

$$384\pi^2 \; R_o^3\mu = \mu \; .$$

With these equations we can now reduce our problem to a Hamiltonian system in two variables β_+, β_- with the Hamiltonian H given by

$$H^2 = p_+^2 + p_-^2 + \frac{3(\mu C)^2\cos^2\gamma}{\sinh^2(2\sqrt{3}\beta_-)} + \frac{3(\mu C)^2\sin^2\gamma\cos^2\lambda}{\sinh^2(3\beta_+ + \sqrt{3}\beta_-)}$$

$$+ \frac{3(\mu C)^2\sin^2\gamma\sin^2\lambda}{\sinh^2(3\beta_+ - \sqrt{3}\beta_-)} + e^{-4\Omega}(V-1) \tag{3.91}$$

$$+ e^{-3\Omega} \mu (1 + (2C)^2 e^{2\Omega} [\sin^2\gamma \sin^2\lambda \ e^{-2(\beta_+ + \sqrt{3}\beta_-)}$$

$$+ \sin^2\gamma \cos^2\lambda \ e^{-2(\beta_+ - \sqrt{3}\beta_-)}$$

$$+ \cos^2\gamma \ e^{4\beta_+}])^{1/2} \quad . \tag{3.92}$$

The functions λ and γ are functions of Ω only given by rewriting Eq.(3.87) with the help of Eq.(3.89) to give

$$\dot{\lambda} = \frac{8Ce^{-\Omega}}{H} \frac{\cos\gamma e^{(\beta_+ + \sqrt{3}\beta_-)}\sinh(3\beta_+ - \sqrt{3}\beta_-)}{(1 + D)^{1/2}} + \frac{8Ce^{-\Omega}}{H} \frac{\cos\gamma \sin^2\lambda \sinh(2\sqrt{3}\beta_-)}{(1 + D)^{1/2}}$$

$$- \frac{3\mu C}{H} \frac{\cos\gamma}{\sinh^2(2\sqrt{3}\beta_-)} - \frac{3\mu C}{H} \cos\gamma \sin\lambda \cos\lambda \left\{ \frac{1}{\sinh^2(3\beta_+ - \sqrt{3}\beta_-)} - \frac{1}{\sinh^2(3\beta_+ + \sqrt{3}\beta_-)} \right\} ,$$

$$\tag{3.93}$$

$$\dot{\gamma} = \sin\gamma \left\{ \frac{8Ce^{-\Omega}}{H} \frac{\cos\lambda \sin\lambda e^{-2\beta_+}\sinh(2\sqrt{3}\beta_-)}{(1 + D)^{1/2}} + \frac{3\mu C}{H} \frac{\cos^2\lambda}{\sinh^2(3\beta_+ + \sqrt{3}\beta_-)} \right.$$

$$\left. + \frac{3\mu C}{H} \frac{\sin^2\lambda}{\sinh^2(3\beta_+ - \sqrt{3}\beta_-)} \right\} ,$$

$$\tag{3.94}$$

where $D = (\overrightarrow{Rn}) \cdot \overrightarrow{e^{-2\beta d}Rn}$.

The Hamiltonian and the walls which are associated with its various potentials have been discussed before. The only matters left are to characterize the effect of γ and λ on the centrifugal potentials and to see what Eqs. (3.93) (3.94) tell us about the behavior of γ and λ. We shall call the three quantities, $\cos^2\gamma$, $\sin^2\gamma\cos^2\lambda$, and $\sin^2\gamma\sin^2\lambda$ the *transparencies* of their respective centrigugal walls and consider that as these quantities vary from 0 to 1, their respective walls vary from "transparent" to "opaque".

The equations that govern γ and λ are quite complex, but we can recognize that there are regions in which the Ω-derivatives of these quantities will be large and regions where they will be small. After Matzner[15] we shall call the regions where γ and λ are rapidly changing *tumbling* regions and where they change slowly *quiescent* regions.

The locations of the tumbling and quiescent regions are difficult to describe because of the complexity of Eqs. (3.93), (3.94). We may, however, point out that they are associated with "potentials" which resemble the potentials which govern the motion of the universe point. This implies that the quiescent and tumbling regions are associated with the existing walls in the sense that whether or not a point lies in one region or the other depends only on its distance from some existing wall. How those regions vary with distances from particular walls may be seen from Eqs. (3.93), (3.94). Notice that the regions surrounding the axes which determine the centrifugal potentials are all tumbling regions.

Because of the complexity of Eqs. (3.93), (3.94) a diagrammatic solution in the sense of a schematic, pictoral representation of the motion is very difficult, though it should be possible. While, because of this complexity, we shall not attempt such a solution, we can make some general statements about the behavior of the universe, and especially about the singularity.

4. *The Quantum Behavior of the Symmetric and General Cases*

With symmetric and general type IX universes we have reached one of the frontiers of research in quantum cosmology. Without being

able to quantize the matter which we include in our model, we cannot
construct a complete, quantum-mechanical solution to our problem. Even in
the absence of a procedure to quantize relativistic fluid matter it is
enlightening to make some remarks about symmetric type IX universes in
which we treat the matter classically. This case provides us with an
example where we can compare two types of prescriptions for quantization,
one where we impose constraints after quantization, and the full ADM
method where we impose constraints before quantizing.

If we quantize by the full ADM method, we impose the constraint
$p_\phi = \mu C$ in the symmetric case before we quantize and arrive at a Hamiltonian
in β_+ and β_- containing the c-number μC. This Hamiltonian, when we
impose the usual commutation relations, gives (using (3.90))

$$-\frac{\partial^2 \psi}{\partial \Omega^2} + \frac{\partial^2 \psi}{\partial \beta_+^2} + \frac{\partial^2 \psi}{\partial \beta_-^2} + \left\{ \frac{3(\mu C)^2}{\sinh^2(2\sqrt{3}\beta_-)} + e^{-4\Omega}(V(\beta_+,\beta_-)-1) + \mu e^{-3\Omega}(1+2C)^2 e^{2\Omega} e^{4\beta_+})^{1/2} \right\}$$

$$(3.95)$$

In this case Hamilton's equation (see Table III.1) for $\dot\phi$ becomes a simple
operator in Ω, and the c-number μC. The solution for ϕ as an operator is a
complicated exercise in what is meant by the expression $\dot\phi$.

We can quantize by means of the alternate scheme if we allow $\psi = \psi(\phi, \beta_\pm)$
and quantize before we impose the constraints. This is not a straightforward problem. In Section VII we show that the quantum-mechanical
equation should be $\psi^{;A}{}_{;A} = 0$, the covariant d'Alembert equation in a space
with a metric

$$ds^2 = -d\Omega^2 + d\beta_+^2 + d\beta_-^2 + \sinh^2(2\sqrt{3}\beta_-)d\phi^2 \qquad (3.96)$$

This leads to

$$-\frac{\partial^2 \psi}{\partial \Omega^2} + \frac{\partial^2 \psi}{\partial \beta_+^2} + \frac{1}{sinh(2\sqrt{3}\beta_-)} \frac{\partial}{\partial \beta_-} (sinh(2\sqrt{3}\beta_-) \frac{\partial \psi}{\partial \beta_-})$$

$$+ \frac{3 \frac{\partial^2 \psi}{\partial \phi^2}}{sinh^2(2\sqrt{3}\beta_-)} + e^{-4\Omega}(V-1)\psi + \mu e^{-3\Omega}(1+(2C)^2 e^{2\Omega} e^{4\beta_+})^{1/2}\psi = 0 . \quad (3.97)$$

This equation is separable as $\psi = \Phi(\phi)\psi(\beta_\pm,\Omega)$ and we get

$$\frac{d^2\Phi}{d\phi^2} - \epsilon\Phi = 0 \qquad\qquad\qquad (3.98)$$

$$-\frac{\partial^2 \psi}{\partial \Omega^2} + \frac{\partial^2 \psi}{\partial \beta_+^2} + \frac{1}{sinh(2\sqrt{3}\beta_-)} \frac{\partial}{\partial \beta_-} (sinh(2\sqrt{3}\beta_-) \frac{\partial \psi}{\partial \beta_-})$$

$$+ e^{-4\Omega}(V-1)\psi + \mu e^{-3\Omega}(1+(2C)^2 e^{2\Omega} e^{\psi\beta_+})^{1/2}\psi = \frac{3\epsilon\psi}{sinh^2(2\sqrt{3}\beta_-)} \quad (3.99)$$

The first equation gives $\Phi = e^{\sqrt{\epsilon}\phi}$, which is exponential or trigonometric depending on whether ϵ is positive or negative. The prescription of Dirac for constrained Hamiltonians now uses the constraint $p_\phi \psi = \mu C \psi = -i \frac{\partial \psi}{\partial \phi}$ to give $\sqrt{\epsilon}$ imaginary, and $\sqrt{\epsilon} = i\mu C$. The fact that $\sqrt{\epsilon}$ is imaginary is heartening since we want $\phi = 0$ and $\phi = \pi/2$ to represent the same metric. We find however that this requirement gives us $1 = e^{i\mu C \pi/2}$, or $\mu C = 4n$, $n = 0,1,\cdots$ or in usual units $\mu C = 4n\hbar$. This is an inconsistency since one assumed μC to be a c-number. On the other hand, we may take this as evidence that in a fully quantized theory μC would obey such a quantum condition.

Even if we ignore the problem of ϕ, the two methods of quantization we have discussed have two different equations in $\beta_\pm$. This discrepancy will be discussed in Section VII.

Research in this field has not proceeded far enough to be able to resolve the difference between the two methods of quantization in this case. The fact that such a discrepancy exists underscores again the usefulness of Hamiltonian cosmology as a simple model for which complicated questions in general relativity can be tested and various methods compared in such a way that their similarities and differences are easily visible.

We shall not attempt to quantize the general case because of its complexity. We shall only say that in the case where we solve all the constraints classically before we quantize, the flapping and fading of the walls should not cause a wave packet much more trouble than in the symmetric case.

5. *Another Hamiltonian Approach to the Symmetric Case*

Ozsvath[25] has used a Lagrangian formulation for the symmetric case which is due to Gödel[23] to obtain a Hamiltonian formulation of this case. He applies this formulation to the special case of diagonal type IX universes and transforms the resultant equations into an analytic system. He conjectures that a similar procedure of *regularization* will lead to a solution for symmetric type IX universes.

F. Other Bianchi Types

At present the Hamiltonian formalism has only begun to be applied to Bianchi types other than I and IX. The Lagrangian formalism has been applied by Hawking[14] and Matzner[48] to Bianchi type V universes. The only investigation of other Bianchi types in the Hamiltonian formulation has been carried out by Jacobs and Hughston[26]. Because their work is still not complete, we shall only give a survey of their current results. They have considered all Bianchi types in which β_{ij} is a diagonal matrix. They have considered matter Lagrangians for pure magnetic fields and comoving fluids. Magnetic field Lagrangians are non-zero only for certain universes as are pointed out in Section III C. Jacobs and Hughston allow combinations of fluids with equations of state $p = (\gamma_n - 1)\rho$, whose Lagrangians have $L_M^i = 0$, $L_M^o = 16\pi\mu R_o^{3(1-\gamma)} e^{3(\gamma-1)\Omega}$. The choice of $L_M^i = 0$ does not in any case seem to be impossible although in some cases it reduces the number of degrees of freedom of the gravitational field. We shall consider their results in two sections, classical and quantum. We shall include their results for types I and IX when they are illuminating.

Classical Behavior. Using the fact that β_{ij} is diagonal and the fact that ^3R can be written as $\frac{1}{3} e^{-4\Omega} V(\beta_+, \beta_-)$ we can use Eqs. (2.19)(3.40) to find

$$H^2 = p_+^2 + p_-^2 + 12\pi^2 R_o^4 e^{-4\Omega}(V) + 24\pi^2 g^{1/2} L_M^o \qquad (3.100)$$

(Note that this parameterization of ^3R varies slightly from that of Misner.) To begin with we consider only perfect fluids for the matter Lagrangian. This means that L_M^o is a function of Ω only.

Jacobs and Hughston[26] compute the space constraints for all of the Bianchi types. Table III.2 gives the potentials associated with each Bianchi type and the space constraints $C^i = 0$ reduced to a statement about constraints on the momenta. The potentials are exponential in every case, so they can be replaced by walls as in the type IX case. These walls move in various directions and at various constant velocities depending on the particular Bianchi type. These walls are affected by changes in H as the ones in the type IX case are. Figure (3.9) shows the walls associated with the potentials and arrows give their velocities. We shall discuss the potentials associated with the magnetic-field solutions later.

In these diagonal Bianchi cases the ADM procedure must be carried out fully, that is, the constraints must be solved and substituted into the Hamiltonian before solving Hamilton's equations. In type V, for instance, the constraint $p_+ = 0$, $\beta_+ = const$, if applied after solving Hamilton's equations, would reduce most general solutions to a set of disjoint points. For details of how these constraints are applied in specific cases, see Jacobs and Hughston[26].

TABLE III.2

I	0	*None*
II	$e^{-8\beta_+}$	*None*
III	$4e^{-(2\beta_+ - 2\sqrt{3}\beta_-)}$	$p_- = 0$
IV	$e^{4\beta_+}(12 + e^{4\sqrt{3}\beta_-})$	$p_+ = 0$
V	$12e^{4\beta_+}$	$p_+ = 0$
VI $(h \neq 0,1)$	$4(1 + h + h^2)e^{4\beta_+}$	$p_+ = \sqrt{3}\ \dfrac{(h+1)}{(h-1)}\ p_+$
VII $h^2 < 4$	$2e^{4\beta_+}[\cosh(4\sqrt{3}\beta_-) + (2h^2-1)]$	*None* $(h=0)$ $p_- = \sqrt{3}p_+ (h \neq 0)$
VIII	$e^{-8\beta_+} + 2e^{4\beta_+}[\cosh(4\sqrt{3}\beta_-)-1] + 4e^{-2\beta_+}\cosh(2\sqrt{3}\beta_-)$	*None*
IX	$e^{-8\beta_+} + 2e^{4\beta_+}[\cosh(4\sqrt{3}\beta_-)-1] - 4e^{-2\beta_+}\cosh(2\sqrt{3}\beta_-)$	*None*

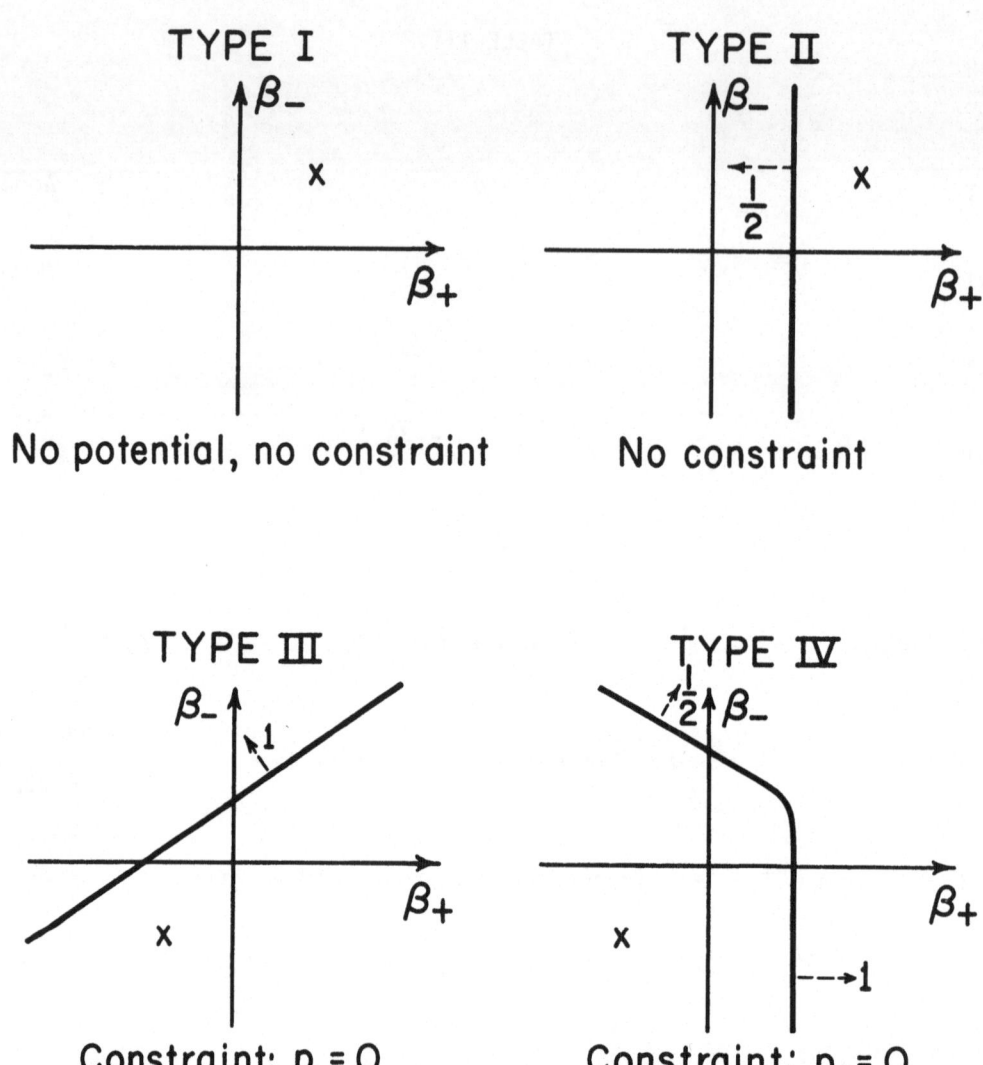

Figure 3.9.

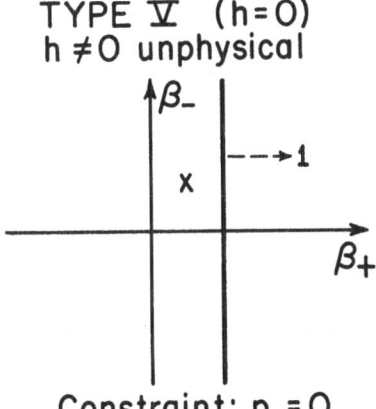

TYPE Ⅴ (h=0)
h ≠ 0 unphysical

Constraint: $p_+ = 0$

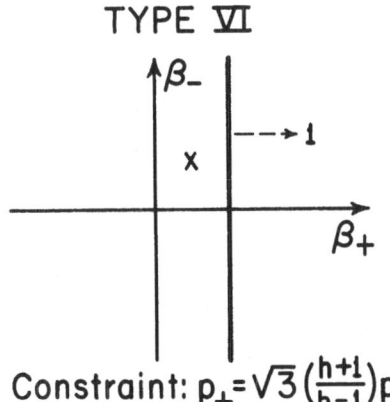

TYPE Ⅵ

Constraint: $p_+ = \sqrt{3}\left(\frac{h+1}{h-1}\right)p_+$

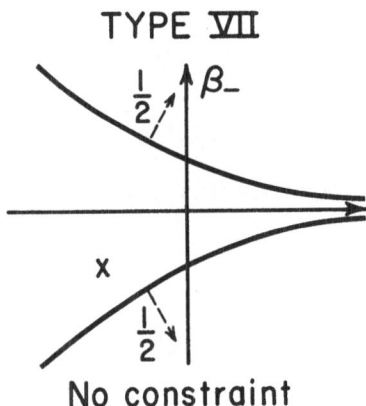

TYPE Ⅶ

No constraint

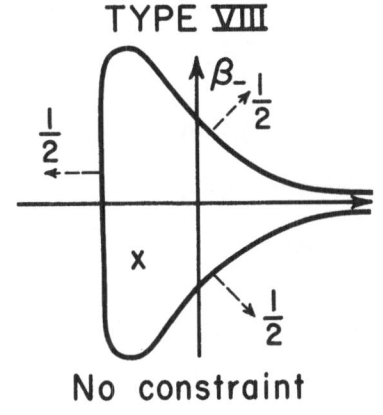

TYPE Ⅷ

No constraint

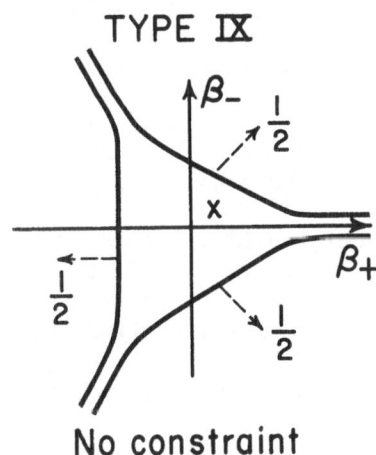

TYPE Ⅸ

No constraint

Figure 3.9. The walls associated with the potentials in diagonal Bianchi-type cosmologies. The heavy lines represent represent the walls and their associated arrows and numbers the magnitude and direction of their velocities (in the sense of $d\beta_{wall}/d\Omega$). The symbol × marks the position of a generic universe point.

The only salient feature which appears in these universes which is not shown by bounces from the walls occurs in type VIII. The potential for type VIII does not have a stable minimum point as the type IX potential does at $\beta_+ = \beta_- = 0$. The universe point for such universes always has a tendency to drift toward positive β_+.

Quantum Behavior. Jacobs and Hughston quantize the Hamiltonian (3.100) by the ADM method, that is, they solve the constraints classically, then quantize the minimal system. They do this by the usual SKG prescription, $H^2 \rightarrow -\dfrac{\partial^2}{\partial \Omega^2}$, $p_\pm^2 \rightarrow -\dfrac{\partial^2}{\partial \beta_\pm^2}$. Because of the fact that superspace for diagonal metrics is flat (see section VI), writing $H^2 = 0$ as a covariant d'Alembertian leads to just such a prescription. We find for our wave equation

$$-\frac{\partial^2 \psi}{\partial \Omega^2} + \frac{\partial^2 \psi}{\partial \beta_+^2} + \frac{\partial^2 \psi}{\partial \beta_-^2} + \{12\pi^2 R_o^4 \, e^{4\Omega}(V(\beta_+, \beta_-)) + 24\pi^2 g^{1/2} L_M^o(\Omega)\}\psi = 0. \quad (3.101)$$

Exact solutions have not been obtained in all cases, but in Table III.3 we give those that have for various L_M^o possibilities. The cases for which analytic solutions do not seem possible are susceptible to approximate analyses of the type used for type IX universes, because all of the potentials are built out of expanding exponential walls.

Magnetic-Field Solutions. Magnetic fields can exist only in types I, II, III, VI ($h=-1$) and VII ($h=0$). In the analysis of the quantum problem by Jacobs and Hughston the electromagnetic quantities are not quantized and are treated as c-numbers. Jacobs and Hughston have computed L_M^o for all of these cases and find

TABLE III.3

BIANCHI TYPE	EQUATION	VACUUM SOLUTION[a,b]	FLUID MATTER SOLUTION
I	$\Box\psi + \sum\limits_n \mu_n e^{3(\gamma_n-2)\Omega} = 0$	$\psi \propto exp[i(p_+\beta_+ + p_-\beta_- \pm\sqrt{p_+^2+p_-^2}\,\Omega)]$	1 component: $exp[i(p_+\beta_+ + p_-\beta_-)] \times$ $\psi \propto Z_{\left[\frac{2ik}{3(2-\gamma)}\right]}\left(\frac{-2\sqrt{\mu}}{3(2-\gamma)} e^{-3(2-\gamma)\Omega/2}\right)$ $(1\le\gamma\le2)$ $exp[i(p_+\beta_+ + p_-\beta_- \pm\sqrt{p_+^2+p_-^2}\,\Omega)]\gamma = 2$ 2 components: Only soluble for $\gamma_2 = 2(\gamma_1-1)$ where: $\psi \propto exp[i(p_+\beta_+ + p_-\beta_- - \frac{3}{2}(\gamma_1-2)\Omega] \times$ $W_{\left[\frac{-\mu_1}{6\sqrt{\mu_2}(\gamma_1-2)}\right],\left[\frac{(p_+^2+p_-^2)^{\frac{1}{2}}}{3(\gamma_1-2)}\right]}\left(\frac{3\sqrt{\mu_2}}{3(\gamma_1-2)} e^{3(\gamma_1-2)\Omega}\right)$
II	$\Box\psi + \frac{1}{3}e^{-4(\Omega+\beta_+)}\psi$ $+ \sum\limits_n \mu_n e^{3(\gamma_n-2)\Omega}\psi = 0$	$\psi \propto exp\{i[p_-\beta_- + \frac{\sqrt{3}}{3}(2\Omega+\beta_+)]\} \times$ $Z_{\left[\frac{p_-^2-p_+^2}{12}\right]}\left(\frac{i}{6}e^{-2(2\beta_++\Omega)}\right)$	None yet obtained.

a. Z is a Bessel function.

b. W_{ab} is a Whittaker function of order a,b.

TABLE III.3 (Continued)

BIANCHI TYPE	EQUATION	VACUUM SOLUTION	FLUID MATTER SOLUTION
III	$$\frac{\partial^2 \psi}{\partial \Omega^2} - \frac{\partial^2 \psi}{\partial \beta_+^2} + \frac{4}{3} e^{-2(\beta_+ + 2\Omega)} \psi$$ $$+ \sum_n \mu_n e^{3(\gamma_n - 2)\Omega} \psi = 0$$	$$\psi \propto exp\{ip_- \frac{\sqrt{3}}{3}(2\beta_+ + \Omega)\} \times$$ $$Z_{\left[\frac{ip_-}{\sqrt{3}}\right]}\left[\frac{2}{3} e^{-(2\Omega + \beta_+)}\right]$$	None yet obtained.
IV	$$\frac{\partial^2 \psi}{\partial \Omega^2} - \frac{\partial^2 \psi}{\partial \beta_-^2} + \frac{1}{3} e^{-4\Omega}(12 + e^{4\sqrt{3}\beta_-}) \psi$$ $$+ \sum_n \mu_n e^{3(\gamma_n - 2)\Omega} \psi = 0$$	None yet obtained.	None yet obtained.
V	$$\frac{\partial^2 \psi}{\partial \Omega^2} - \frac{\partial^2 \psi}{\partial \beta_-^2} + 4 e^{-4\Omega} \psi$$ $$+ \sum_n \mu_n e^{3(\gamma_n - 2)\Omega} \psi = 0$$	$$\psi \propto e^{ip_- \beta_-} Z_{\left[\frac{ip_-}{2}\right]}[e^{-2\Omega}]$$	1 component, ($\gamma = 4/3$): $$exp(ip_- \beta_- + \Omega) \times$$ $$W_{\left[-\frac{\mu_r}{8}\right], \left[\frac{p_-}{2}\right]}[2e^{-2\Omega}]$$

TABLE III.3 (Continued)

BIANCHI TYPE	EQUATION	VACUUM SOLUTION	FLUID MATTER SOLUTION
VI	$\dfrac{\partial^2 \psi}{\partial \Omega^2} - \dfrac{\partial^2 \psi}{\partial \beta_-^2} + \dfrac{4}{3} e^{-4(\Omega-\beta_+)}\psi$ $+ \sum_n \mu_n e^{3(\gamma_n-2)}\psi = 0$	$\psi \propto exp[(12)^{-\frac{1}{2}} R\{cos\phi[e^{-4(\Omega-\beta_+)}$ $+ R^{-2}(\Omega+\beta_+)] + i\,sin\phi[e^{-4(\Omega-\beta_+)}$ $- R^{-2}(\Omega+\beta_+)]\}]$, where $Re^{i\phi}$ is a constant restricted only by $cos\phi<0$	None yet obtained.
VII	$\Box\psi + \dfrac{1}{3} e^{-4\Omega} V_{VII}(\beta_\pm)\psi$ $+ \sum_n \mu_n e^{3(\gamma_n-2)\Omega}\psi = 0$	None yet obtained.	None yet obtained.
VIII	$\Box\psi + \dfrac{1}{3} e^{-4\Omega} V_{VIII}(\beta_\pm)\psi$ $+ \sum_n \mu_n e^{3(\gamma_n-2)\Omega}\psi = 0$	None yet obtained.	None yet obtained.
IX	$\Box\psi + \dfrac{1}{3} e^{-4\Omega} V_{IX}(\beta_\pm)\psi$ $+ \sum_n \mu_n e^{3(\gamma_n-2)\Omega}\psi = 0$	None yet obtained.	None yet obtained.

I: $-B^2 e^{\Omega} (e^{+2\beta_+ + 2\sqrt{3}\beta_-}$ or $e^{2\beta_+ - 2\sqrt{3}\beta_-}$ or $e^{-4\beta_+})$

II: $-B^2 e^{\Omega} (e^{+2\beta_+ + 2\sqrt{3}\beta_-}$ or $e^{2\beta_+ - 2\sqrt{3}\beta_-})$ (3.102)

III, VI $(h=-1)$, VII $(h=0)$: $-B^2 e^{\Omega} e^{-4\beta_+}$

where B is a constant. Figure (3.10) shows the walls associated with these potentials, the arrows give their velocities. These walls are meant to be superposed on the walls already appearing from the V term in these cases. Handling the addition of extra walls in the classical problem is familiar from the various type IX cases and should present no problem here.

The only case in which a quantum solution has been obtained is that of the type I universe with a magnetic wall. By symmetry we can take the wall perpendicular to the β_+-axis. The Klein-Gordon equation is

$$- \frac{\partial^2 \psi}{\partial \Omega^2} + \frac{\partial^2 \psi}{\partial \beta_+^2} + \frac{\partial^2 \psi}{\partial \beta_-^2} + 24\pi^2 B^2 e^{-2\Omega} e^{-4\beta_+} \psi = 0,$$ (3.103)

for which Jacobs and Hughston give the solution

$$\psi \propto exp[i\{m\beta_- + (k\sqrt{3})(2\Omega + \beta_+)\}] Z_{\left[\frac{m^2-k^2}{3}\right]} \left[\frac{iB}{\sqrt{3}} e^{-2(\beta_+ + \Omega)}\right] .$$ (3.104)

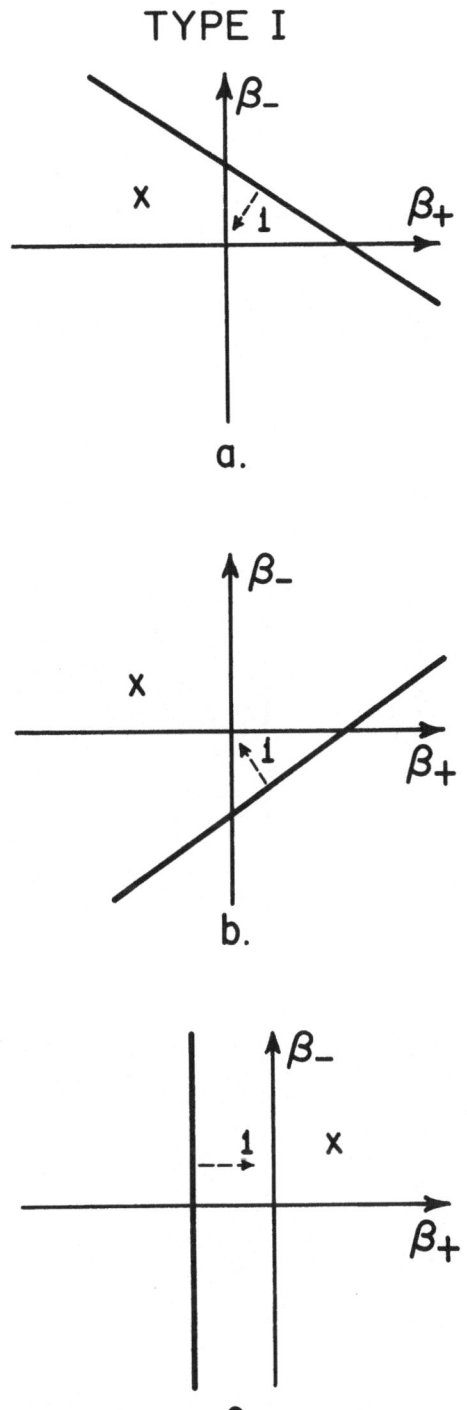

Figure 3.10.

TYPE Ⅱ

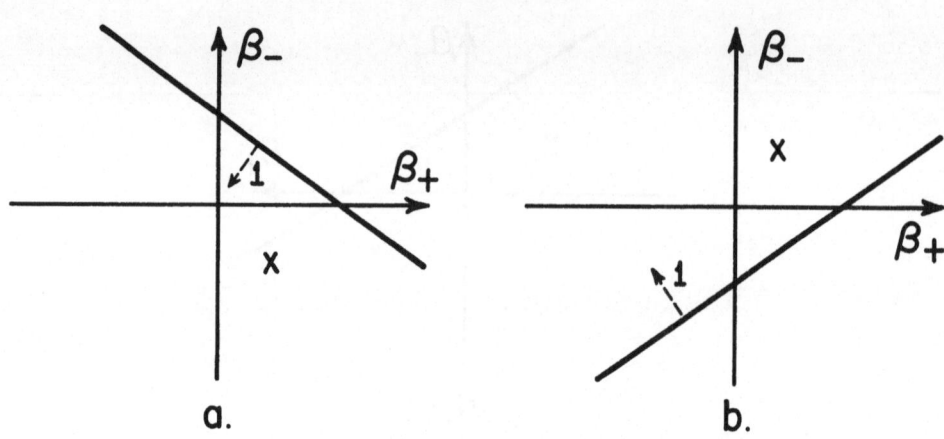

a.   b.

TYPES Ⅲ, Ⅵ (h=-1), Ⅶ (h=0)

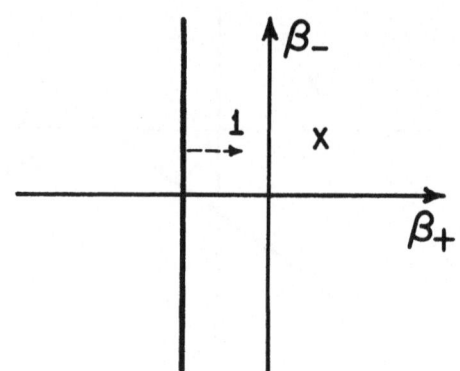

Figure 3.10. The walls associated with the magnetic potentials. The
meaning of the arrows and numbers is the same as in
Figure (3.9). The letters a,b, etc. label different
possible magnetic walls for each Bianchi type.

IV. THE HAMILTONIAN FORMULATION APPLIED TO MORE
COMPLEX SYSTEMS

Eventually in our study of Hamiltonian cosmology we shall want to proceed from the simple homogeneous cosmologies to inhomogeneous ones in order to extend our understanding of such problems as the nature of the singularity for the classical case and to see if quantization will lead to elimination of the singularity. We might hope that we can find inhomogeneous cosmologies which will still have a cosmic time, that is, for which we can write,

$$ds^2 = -dt^2 + g_{ij}(x,t)dx^i dx^j \ . \tag{4.1}$$

Indeed at least one such cosmology has been found by Belinskii and Khalatnikov[17].

At this point there has been no investigation in the Hamiltonian formulation of any inhomogeneous cosmologies. There have, however, been several studies of well-known metrics of general relativity which have inhomogeneous curvature of their space sections. In this section we present two of these as examples of ways to handle such problems in the hope that they will indicate the scope of the problem of inhomogeneous cosmologies and perhaps point the way toward its solution.

The feature of the problem of inhomogeneous three-spaces which will be most outstanding in the study of inhomogeneous cosmologies is that the Hamiltonian becomes a Hamiltonian density, and we must deal not with a particle problem, but with a field theory. This will become clear as we proceed.

The Einstein-Rosen Metric by the Kuchař Method

Kuchař[29] has considered the metric of Einstein and Rosen[49] for cylindrical waves.   He writes the metric as

$$ds^2 = e^{2(\gamma-\psi)}(dr^2-dt^2) + r^2e^{-2\psi}d\phi^2 + e^{2\psi}dz^2 , \qquad (4.2)$$

when ψ and γ are functions of r and t. The functions ψ and γ obey the equations

$$\frac{\partial^2\psi}{\partial r^2} + \frac{1}{r}\frac{\partial\psi}{\partial r} - \frac{\partial^2\psi}{\partial t^2} = 0, \qquad (4.3a)$$

$$\frac{\partial\gamma}{\partial r} = r\left[(\frac{\partial\psi}{\partial r})^2 + (\frac{\partial\psi}{\partial t})^2\right] , \qquad (4.3b)$$

$$\frac{\partial\gamma}{\partial t} = 2r\,\frac{\partial\psi}{\partial r}\,\frac{\partial\psi}{\partial t} . \qquad (4.3c)$$

In order to apply the ADM method Kuchař writes the space part of this metric as

$$ds^2 = e^{2(\gamma-\psi)}dr^2 + e^{-2\psi+2\ell n\lambda}d\phi^2 + e^{2\psi}dz^2 . \qquad (4.4)$$

He constructs π^{ij} by means of the following prescription: First, let

$$\pi^{ij} = diag(\frac{1}{2}\,p_r e^{-2(\gamma-\psi)},\ \frac{1}{2}\,p_\phi e^{2(\psi-\ell n\lambda)},\ \frac{1}{2}\,p_z e^{-2\psi}) . \qquad (4.5)$$

Second, construct $\pi^{ij}\dot{g}_{ij} = \dot{\psi}(-p_r-p_\phi+p_z) + \dot{\gamma}p_r + \dot{\lambda}(p_\phi/\lambda)$ and by defining

$$\pi_\psi = -p_r - p_\phi + p_z \qquad (4.6a)$$

$$\pi_\lambda = p_\phi/\lambda \qquad (4.6b)$$

$$\pi_\gamma = p_r , \qquad (4.6c)$$

arrive at

$$\pi^{ij}\dot{g}_{ij} = \pi_\psi\dot{\psi} + \pi_\lambda\dot{\lambda} + \pi_\gamma\dot{\gamma} , \tag{4.7}$$

and

$$\pi^{ij} = diag(\frac{1}{2}\pi_\gamma e^{-2(\gamma-\psi)}, \frac{1}{2}\lambda\pi_\lambda e^{2(\psi-\ell n\,\lambda)}, \frac{1}{2}(\pi_\psi+\pi_\gamma+\lambda\pi_\lambda)e^{-2\psi}). \tag{4.8}$$

From (4.4) we see that this choice puts $\pi^{ij}\dot{g}_{ij}$ into canonical form with respect to the metric parameters ψ, γ, λ and their conjugate momenta π_ψ, π_λ, π_γ.

For the rest of the ADM formalism we need only compute C^0 and C^i, and solve them in concert with our choice of coordinate conditions. Computations of C^0 and C^i are given by Kuchař, who finds

$$C^0 = e^{\psi-\gamma}[\frac{1}{8}\lambda^{-1}\pi_\psi - \frac{1}{2}\pi_\lambda\pi_\gamma + 2(\lambda''-\gamma'\lambda'+\lambda(\psi')^2)] , \tag{4.9a}$$

$$C^1 = e^{2(\psi-\gamma)}[-\pi_\gamma' + (\gamma'-\psi')\pi_\gamma + \lambda'\pi_\lambda + \psi'(\pi_\psi+\pi_\gamma)] , \tag{4.9b}$$

$$C^i = 0, \quad i = 2,3, \tag{4.9c}$$

where prime indicates differentiation with respect to r.

We must now begin to choose coordinate conditions. Kuchař wishes to recover the original Einstein-Rosen form for the metric, so he chooses $\lambda = r$ and $N = e^{\gamma-\psi}$, $N_i = 0$. Here, for the first time, we come across the problem mentioned in Section II of a possible conflict between choosing coordinates by giving them explicitly and giving them by fixing N and N_i. Kuchar checks this consistency of his choice by computing the Einstein equations and comparing them with the equations obtained by varying

the term $NC^0 + N_i C^i$ in the Hamiltonian. The condition for this consistency is found to be

$$\pi_\gamma = 0 \ . \tag{4.10}$$

Kuchař considers this as displaying explicitly the choice of time
coordinate implicit in the choice $N = e^{\gamma - \psi}$, $N_i = 0$. That is, he writes

$$t = - \frac{1}{2} \int \pi_\gamma dr. \tag{4.11}$$

With the above coordinate conditions we find that the constraints
reduce to

$$\gamma' = \frac{1}{16} r^{-1} \pi_\psi^2 + r(\psi')^2 \ , \tag{4.12}$$

$$\pi_\lambda = - \psi' \pi_\psi \ . \tag{4.13}$$

We must now write the action in terms of our new coordinates and
insert the constraints. The action is

$$I = \int (\pi_\psi \dot\psi + \pi_\gamma \dot\gamma + \pi_\lambda \dot\lambda) dr dt,$$

where the insertion of the constraints is implied. We rewrite I as

$$I = \int [\pi_\psi \dot\psi + \frac{d}{dt}(\pi_\gamma \gamma) - \frac{d}{dr}(\gamma \int \dot\pi_\gamma dr) + \gamma'(\int \pi_\gamma dr)^\cdot + \dot\lambda \pi_\lambda] dt dr \tag{4.14}$$

and hence as

$$I = \int [\pi_\psi d\psi - [2(\frac{1}{16} \lambda^{-1} \pi_\psi^2 + \lambda(\frac{d\psi}{d\lambda})^2] d(- \frac{1}{2} \int \pi_\gamma d\lambda) + [-\psi' \pi_\psi] d\lambda] d\lambda \tag{4.15}$$

$$= \int [\pi_\psi d\psi - H \ dt] dr \ ,$$

where $\quad H = 2[\frac{1}{16} r^{-1}\pi_\psi^2 + r(\psi')^2]$.

From the form of I we see that Eq. (4.3a) for ψ follows. The definition of γ' in terms of (4.12) gives γ once ψ has been determined.

If we consider this approach to the Hamiltonian formulation, we see that the two new interesting features are: 1) the use of λ as a general function and its specialization to the desired form by means of coordinate conditions; and 2) the coordinate choice $t = -\frac{1}{2}\int \pi_\gamma dr$.

The Schwarzschild Solution by the Kuchař Method

Nutku and Kobre[30] have applied the Kuchař method to the Schwarzschild problem. They do this by writing the usual Schwarzschild metric as

$$ds^2 = -N^2(\tilde{\omega}^0)^2 + e^{2\mu}(\tilde{\omega}^1)^2 + e^{2\lambda}[(\tilde{\omega}^2)^2 + (\tilde{\omega}^3)^2] , \qquad (4.16)$$

where the forms $\tilde{\omega}^\mu$ are $\tilde{\omega}^0 = dt$, $\tilde{\omega}^1 = dr$ $\tilde{\omega}^2 = d\theta$, $\tilde{\omega}^3 = sin\theta d\phi$. The functions μ, λ and N are taken to be functions of r and t only. In order to parametize $\pi^{ij}\dot{g}_{ij}$, Nutku and Kobre[30] write

$$\pi^{ij} \equiv diag \ (\frac{1}{2} \pi_\mu e^{-2\mu}, \ \frac{1}{4} \pi_\lambda e^{-2\lambda}, \ \frac{1}{4} \pi_\lambda e^{-2\lambda}), \qquad (4.17)$$

to obtain $\pi^{ij}\dot{g}_{ij} = \pi_\mu\dot\mu + \pi_\lambda\dot\lambda$. They give the constraints as

$$C^0 = e^{-\mu-2\lambda}\{\frac{1}{8} \pi_\mu^2 - \frac{1}{4} \pi_\mu\pi_\lambda + 2e^{4\lambda}[2\lambda''-2\lambda'\mu'+3(\lambda')^2-e^{2(\mu-\lambda)}]\} = 0 \quad (4.18)$$

$$C^1 = -e^{-2\mu}(\pi_\mu'-\mu'\pi_\mu-\lambda'\pi_\lambda) = 0 \qquad (4.19)$$

$$C^2 = C^3 = 0, \qquad (4.20)$$

where dot and prime mean the same as previously.

We must now choose coordinate conditions. Nutku and Kobre[30] choose $\lambda = \ell n \, r$ to obtain the usual Schwarzschild metric and, following the choice of Kuchař, $t = -\frac{1}{2} \int \pi_\mu dr$ (or $\pi_\mu = 0$). With these choices $C^1 = 0$ implies $\pi_\lambda = 0$ and $C^0 = 0$ reduces to

$$2r\mu' + e^{2\mu} = 1. \qquad (4.21)$$

This equation has the well-known solution $\mu = -\frac{1}{2} \ell n (1 - \frac{2m}{r})$, so we recover a portion of the usual Schwarzschild solution.

Nutku and Kobre[30] now show that we can write

$$I = \int \mu' \left(\int \pi_\mu dr \right)^{\cdot} dr dt, \qquad (4.22)$$

or in terms of our new coordinates

$$I = - \int 2\mu' dr dt. \qquad (4.23)$$

This form of the action shows that we have a degenerate problem, because no dynamical terms appear in the action. Solving the constraints, as in the Friedmann-Robertson-Walker problem, has exhausted all the dynamical variables. The only term that appears in I is the Hamiltonian, derived from a Hamiltonian density

$$H = 2\mu'.$$

In order to determine if we have actually recovered the Schwarzschild solution we must compute N and N_i from the Einstein equations. If we choose $N_i = 0$, we have $\dot{\pi}_\mu = \delta(NC^0/\delta\mu)$ which gives

$$\dot{\pi}_\mu = 0 = N(1 - e^{2\mu}) + 2r \frac{dN}{dr}$$

A short calculation shows that $N = e^{-\mu}$ satisfies this equation for $\mu = -\frac{1}{2}\ln(1 - \frac{2m}{r})$. We have indeed recovered the Schwarzschild solution.

As a final note, Nutku and Kobre[30] have managed to obtain a non-degenerate Hamiltonian problem by inserting matter in the form of a scalar field. We shall not discuss this here.

V. APPLICATIONS TO BIANCHI-TYPE UNIVERSES

In discussing the applications of Hamiltonian cosmology in this section we shall leave out a detailed discussion of what is perhaps the most important use to which this formulation can be put, that is in quantizing model universes. Because this use is so important, we shall defer discussing it to a later section. In purely classical terms the technique is very useful in providing exact or approximate solutions to complicated cosmologies which allow one to investigate certain general questions.

A. Motions of Homogeneous Universes

The power of the Hamiltonian approach to give easily exact solutions in such cases as the Bianchi-type I universes and the Kantowski-Sachs universe and approximate solutions in the Bianchi-type IX cases and other Bianchi types, and the pictorial nature of the presentation of these solutions enables one to grasp many important features of their time development at a glance. With these solutions one can examine such problems as: 1) The relation between the anisotropy matrix β_{ij} and the anisotropy of the temperature of the 3°K black-body radiation, 2) The effect of rotation on the anisotropy of the black-body radiation, and 3) The nature of the singularities of homogeneous cosmologies. The last of these is important enough that we shall leave it to a separate subsection. The problem of the motion of a homogeneous universe has recently been tied to problems 1) and 2) in computations of the development of anisotropy and rotation by means of observations of the

3°K black-body radiation which give small upper bounds on the present amount of anisotropy and rotation[14] in the universe. We shall discuss this in terms of two of the Bianchi types.

1. *Bianchi-type I.*

In Bianchi type I universes, for instance, we may choose $\beta_{ij} = 0$ at the present time. This fixes a point on the $\beta_+\beta_-$-plane, but gives us no way of choosing which of the infinity of straight lines through the origin represents the actual track of the universe. In order to complete our description of the motion of the universe we need to know R_o, p_+, p_-, H, (and μ and Γ for matter-and-radiation-filled universes). These six quantities are usually determined by observation. Because of the dependence of N on Ω and H and the fact that $dt = -Nd\Omega$, H at the present time is related to the Hubble constant, that is, the Hubble constant gives H/R_o now, if we know Ω now. It is perhaps most convenient to choose some arbitrary value for Ω at present. This choice leaves measurements of the Hubble constant and the deceleration parameter q_o, or the present density of matter to determine H/R_o and μ respectively. The total integrated energy density of the 3°K black-body radiation gives Γ. With these values determined we only need $p_\pm$ (because of the definition of H in terms of $p_\pm$, $\beta_\pm$, R_o, Ω, μ and Γ) to give a complete set of initial data for the universe. Since $d\beta_\pm/dt = p_\pm e^{3\Omega}/12\pi R_o^3$, measurements of the anisotropy in the Hubble constant (whose relation to $d\beta_\pm/dt$ has been pointed out, by Misner[7]) give this last item we need. The discovery of the 3°K black-body radiation and the low limits on its anisotropy[50,51] provides another independent limit on $d\beta_\pm/dt$ and hence on $p_\pm$.

If $\mu = \Gamma = 0$, the universe point must move on the "light cone" in $\beta_+\beta_-\Omega$-space (the cone at 45° to the line $\beta_+ = \beta_- = 0$). The particular line of motion on the cone is determined by p_+/p_-. If μ or Γ is not zero motion inside the light cone is allowed, in particular motion along $\beta_+ = \beta_- = 0$ (Friedmann-Robertson-Walker $k = 0$ universe). The plane of motion is still determined by p_+/p_-.

Misner[7,16] has discussed a way of allowing β_+ and β_- to have been larger in the distant past than this discussion we have given would indicate, by allowing collisionless neutrinos to reduce anisotropy sharply at a high temperature epoch.

In the Bianchi type I case, the Hamiltonian approach gives us an uncomplicated solution (if we allow only fluid matter and radiation). We are able, with a few observationally determined parameters to give the entire motion of the universe.

2. *Bianchi-type IX Universes*

As can be seen from the form of H from the previous section, the problem of type IX universes seems insoluble analytically. It is in such a case that the power of the Hamiltonian approach for classical problems becomes most evident. Because we have been able to reduce the problem to the equivalent of a particle problem in two dimensions for all three cases of Bianchi-type IX universes, we can discuss the solution qualitatively and can give a schematic representation of the motion which gives all of the large-scale features of the motion. A program for doing this was given by Ryan[12], who calls it *qualitative cosmology*. Such a program depends on the fact that the walls of the potentials in the type IX Hamiltonians are all steep enough that we can replace them by infinitely hard walls which expand in Ω. As was mentioned in a previous section, the fact that $\dot{H} \cong 0$ for much of the motion makes this possible.

The fact that when $\dot{H} = 0$ the universe point moves in a straight line enables us to build up a solution for a type IX universe out of straight line segments attached at their ends. These segments must, of course be bounded by the walls associated with the particular type IX case (diagonal symmetric or general) we are considering. We need to know how to attach the ends of the lines, so we need a law of reflection from the walls.

We need to know how collisions from the walls affect the motion of the walls. This means we need to know how H changes when the universe point bounces off the walls. Finally, we need to know how the overall motion of the walls as $\Omega \to \infty$ is affected by collisions.

The law of reflection and the effect of collisions are exercises in Hamiltonian dynamics for a time-dependent potential. Such calculations are given in Appendix C and have the following results for all three type IX cases:

1) The universe point cannot collide with rotation potentials.

2) Reflection from any centrifugal wall is specular and H does not change when the universe point bounces from them.

3) The only walls which have $\dot{H} \neq 0$ during a bounce are the gravitation walls. They have

    a) A law of reflection: $sin(\theta_{out}) = \dfrac{3 \, sin(\theta_{in})}{5-4 \, cos(\theta_{in})}$

    where θ_{in} and θ_{out} are defined by means of Fig. (5.1); and

    b) A law for the change of H during a bounce, $H_{in} \, sin(\theta_{in}) = H_{out} \, sin(\theta_{out})$, where H_{in} and H_{out} are H before and after collision respectively.

Because, for bounces from a single wall of the gravitation potential,

Hamilton's equations may be solved for H exactly, (see Appendix C)

we can compute $\Delta\Omega$, the time spent in collision with the gravitation wall

as $\Delta\Omega = (H_{out} - H_{in})(d\Omega/dH)_{max}$. The quantity $d\Omega/dH)_{max}$ can be computed

from the exact solution for H.

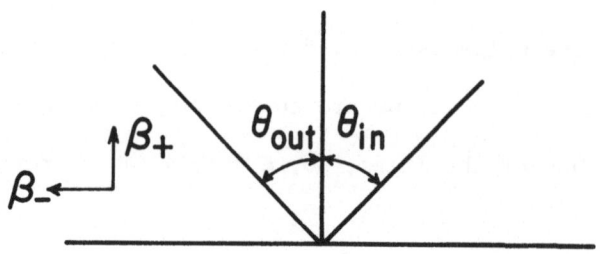

Figure 5.1. The definition of θ_{in} and θ_{out}.

With the quantities given above we can skeletonize the motion of

the walls, having them move with the velocities given in Section IIIE

when $H = 0$, and change their motion instantaneously when the universe

point collides with them. We can differentiate our expressions for the

positions of the walls Eqs.$((3.58, .74-5))$ and replace the terms involving

$dH/d\Omega$ with instantaneous changes given by these expressions with $\frac{\Delta H}{\Delta\Omega}$,

where $\Delta H = (H_{out} - H_{in})$. By means of the procedure we find that for

collisions with the **gravitation walls**,

    1) In all three (diagonal, symmetric and general) cases the

        gravitation walls slow down or stop during collisions;

    2) In the symmetric and general cases the rotation walls step

        in toward the center of the triangular gravitation well during

        collisions;

    3) In the symmetric and general cases the centrifugal walls move

        away from the axes which they surround. In the symmetric case

        this behavior is simple. In the general case the details are

        complicated and depend on the structure of the tumbling regions.

The details of these changes can be worked out from the full expressions for the wall positions (containing H) given in Section IIIE.

As a note, we point out that the adiabatic invariant $H\Omega \approx const$ can be used to test the long-term effect of collisions with the gravitation walls on the motion of all walls. Inserting $H = H_o/\Omega$ into our expressions for the positions of the walls we can see the average behavior of the walls over a large number of collisions. This has been used by Ryan[12] to show that the slowing of the walls due to collisions results on the average only in a logarithmic decrement to be added to the linear expansion, which is not enough to change the conclusions we have reached about the long-term behavior of the walls.

The final problem we need to consider in our qualitative scheme is that of the changes of ϕ, ψ, and θ as the universe point moves throughout the potential wall in the symmetric and general cases. Far from centrifugal walls and tumbling regions, ϕ, ψ, and θ are constant. These angles only change rapidly when the universe point enters a tumbling region, or collides with a centrifugal wall. Therefore we can define $\Delta\phi$, $\Delta\theta$, $\Delta\psi$, the analogues of ΔH for collisions from the gravitation walls, that is, we approximate for our qualitative solution the behavior of the angles by keeping them constant except for abrupt changes when the universe point collides with a centrifugal wall or passes through a tumbling region. In the symmetric case we can define $\Delta\phi$ easily by integrating $\dot{\phi}$ (Table III.1) along the path of the universe point as it collides with the centrifugal wall. The integral is

$$\Delta\phi = 2 \int_{\beta_{min}}^{\infty} \frac{3\mu C \, d\beta_-}{\dot{\beta}_- \, H \, sinh^2(2\sqrt{3}\beta_-)} \qquad (5.1)$$

where β_{min} is the minimum value of β_- for the given constants H and p_+. Because $d\beta_-/d\Omega$ goes from positive to negative at β_{min}, β_{min} occurs when $p_- = 0$. From $\frac{d\beta}{d\Omega} = p_-/H$ and $H^2 = p_+^2 + p_-^2 \ 3(\mu C)^2/sinh^2(2\sqrt{3}\beta_-)$ we can solve for p_- (and hence β_{min}) as a function of β_-, p_+, and H and find β_{min} by letting $p_- = 0$. We can compute the integral in terms of constants of motion p_+ and H, where we let $p_+ = H \ sin \ \zeta_{in}$ and we define ζ_{in} in Fig. (5.2). We find

$$\Delta\phi = \frac{1}{2}\left[sin^{-1}\left\{ \frac{H^2 cos^2(\zeta_{in}) - 3(\mu C)^2}{H^2 cos^2(\zeta_{in}) + 3(\mu C)^2} \right\} + \frac{\pi}{2} \right]. \tag{5.2}$$

This varies between $\pi/2$ and zero as $H \to 0$ (as we approach the singularity).

The problem for the general case is more complicated because of the large number of angles, walls, and regions involved. Expressions similar to $\Delta\phi$ should be possible, however, for the other two angles.

With the above prescriptions for constructing an approximate solution we can begin at the present, with values of the relevant parameters deduced from observational data in much the same way as was done in the Bianchi type I case. We can see in four cases a progressing complexity from a case where the Hamiltonian approach gives a complete solution (Bianchi type I) through two cases where it easily gives detailed approximate solutions (Bianchi type IX, diagonal and symmetric cases) to a case where a confusion of walls and regions overwhelms the simplifying power of the approach to a point where numerical methods would be needed even to obtain a good approximate solution. Notice, however, that the pictorial nature of the presentation of the problem makes the most vital elements of the development of the universe stand out clearly, even in cases where a diagrammatic solution is difficult to obtain. This points out that visual representation of complicated systems is perhaps the strongest feature of the Hamiltonian approach applied to classical problems.

As a final remark, note that we have neglected the fact that the region of the gravitation potential in the type IX cases near $\beta_+ = \beta_- = 0$ has circular rather than triangular symmetry. Needless to say, care must be exercised in our analysis if we wish to assume $\beta_\pm \approx 0$ today. Because of the simplicity of the form of the potential in this region, however, the problem can be studied without approximation easily.

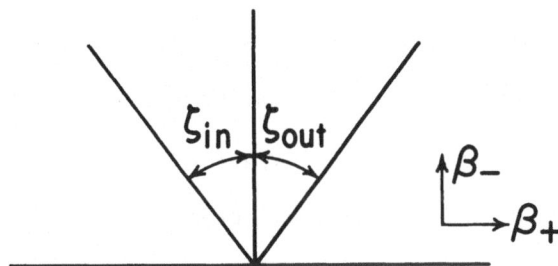

Figure 5.2. Definition of ζ_{in} and ζ_{out}.

B. Singularities

The theorems of Hawking, Penrose, et al.[31] which show that there is a singularity of some sort for every reasonable cosmology depend on showing that if singularities are excised from a cosmological manifold these manifolds are always geodesically incomplete; that is, there exist singularities because if they are removed there are always "holes" in the manifold. Because these theorems are stated in this way they give no clue about what is hidden in the "holes", thus while we know singularities exist we have no information about their character. Until satisfactory theorems are developed to predict this character in general we must investigate each particular case. Hamiltonian cosmology turns out to be an admirable tool for doing this, especially in the case of homogeneous cosmologies. At present the Hamiltonian method has been most completely applied to Bianchi types I and IX and the Kantowski-Sachs universe, so we shall consider them here.

Classically we may use these models to test the effect of various features of the present-day universe on the character of the singularity. For instance, the more complicated type IX universes may be used to test the effect of rotation on the singularity. Raychaudhuri's equation[52] (for perfect fluids with $p=(\gamma-1)\rho$) reads

$$\ddot{R} - \frac{R}{3}\left(\Lambda - \Sigma_{\mu\nu}\Sigma^{\mu\nu} + \Omega_{\mu\nu}\Omega^{\mu\nu}\right) + \mu R^{1-3\gamma} = 0, \qquad (5.3)$$

where, $\Sigma_{\mu\nu}$ and $\Omega_{\mu\nu}$ are the tensors of shear and rotation respectively,[53] Λ is the cosmological constant, and μ is a constant related to matter density, $\Sigma_{\mu\nu}$ and $\Omega_{\mu\nu}$ were defined in Section I. Because of the sign of the $\Omega_{\mu\nu}\Omega^{\mu\nu}$ term, this equation seems to indicate that rotation could have the effect of cancelling the singularity, which we know from the singularity theorems not to be the case. It does, however, hold the possibility of radically changing the nature of the singularity from that of universes which have only expansion and shear. We are able to show below that this is not the case.

The one important possibility that the singularity theorems do not consider is that of quantum general relativity. It has long been thought that perhaps a quantum theory of gravity would contain terms which would prevent a singularity. The models given in the previous section and their quantum behavior can be used to give us a clue about the effect of quantization on the existence of the singularity. Of course, we must stress that the clues they give us are just that, clues. Because we have imposed symmetries before quantization the models we have are only models. A complete description must wait until we are able to quantize the full inhomogeneous problem. If, however, we can trust the clues which the homogeneous models give us, quantum mechanics will be of little help in preventing a singularity.

We shall study both the classical and quantum behavior of the homogeneous models near the singularity, concentrating on Bianchi-types I and IX, the Kantowski-Sachs universe, and the Friedmann-Robertson-Walker universes.

1. *Bianchi Type I Universes*

a. Classical Singularity

Classically, the singularity of Bianchi type I universes which corresponds most closely to the singularities which occur in the

Friedmann universes is that which occurs when $\Omega \rightarrow \infty$. From the equations for fluid matter in these universes we see that this is a true singularity in the sense that matter contained in the universe reaches infinite density at this point.

The solution for these universes is some straight line in $\beta_+\beta_-\Omega$-space which inexorably cuts the $\Omega = +\infty$ surface at some time in its history. While we know from the singularity theorems that this universe must have a singularity, it is the solution which shows us that such a singularity must be of a relatively simple type similar to that of the Friedmann universes.

It is instructive to consider what would have been required for the universe to have avoided this singularity. Two possibilities exist:

1) The solution could have been bounded in the Ω-direction (that is, it could have simply turned around in $\Omega\beta_+\beta_-$-space)

2) The solution could behave in some complex manner indicating a complete breakdown of the solution caused by, perhaps, an attempt by the underlying manifold to change its topology.[54]

Of course, neither of these possibilities occurs; the solution marches peacefully through $\Omega\beta_+\beta_-$-space with no indication of escaping the singularity. In order for the first possibility to occur, we would have had to have H go through zero in order that it might proceed to negative values and change Ω from an increasing to a decreasing function of t. An empty type I universe has $H = const$ while a matter-filled one has $H \sim e^{-3\Omega/2}$, and fluid radiation gives $H \sim e^{-\Omega}$; in the fluid universes H behaves in the same manner as in the Friedmann case. In none of these cases does $H \rightarrow 0$ before the singularity at $\Omega = \infty$.

b. Quantum Mechanical Singularity

If we go over to the quantum model, we see that there is very little that we get that is new. In the $\beta_+\beta_-$-plane a wave packet constructed from the plane-wave solutions of Section IIIA moves sedately along one of the classical trajectories with no indication of suspicious behavior. The only other possibility is that the quantum mechanical solution turns around at some time t. As has been pointed out by Zapolsky[13] the quantity $d\Omega/dt$ is an operator because it is equal to $-\frac{1}{N}$ which is given by $-(12\pi R_o^3)\,e^{3\Omega}H$, a combination of the operator H and the c-number Ω. Because of this, some care must be exercised in applying the words "expanding" and "contracting" to the universe. In the Bianchi type I case, however, it seems straightforward to discriminate between these two states by means of the quantity $sgn\ \ H|\psi>$ or $sgn\ <\psi|H|\psi>$ Both of these remain either + or − for an empty universe, since H operating on a wave packet merely brings down the constant $\sqrt{k_+^2+k_-^2}$ to be integrated over. The expectation value of H will be a constant. If we ignore the problem of quantizing any matter we add, the behavior of H mimics that of the classical H. Because of the points we have mentioned, the singularities of Bianchi type I universes seem not to be affected by quantization. Note also that the universe is neither more nor less "quantum-mechanical" at any time in its development.

We have shown that classically and quantum mechanically type I universes have an infinite-density singularity. We shall now proceed to more complicated universes and see if they present anything new.

2. *Kantowski-Sachs Universe*

a. Classical Singularity

Because the Kantowski-Sachs universe has two singularities, an initial one and one due to recollapse, the universe point crosses the $\Omega_\lambda = +\infty$, (a λ subscript refers to quantities associated with the λ-time formulation (see Section IIIB)) surface in two places. Note, however, that on each approach to this surface it is moving in a straight line in $\Omega_\lambda \beta_\lambda$-space. In these regions the motion is an exact analogue of the motion of the universe point in the Bianchi type I case. We can carry over the analysis of that case to that of the Kantowski-Sachs universe.

In the Kantowski-Sachs case we again see no indication of suspicious behavior in the $\Omega_\lambda \beta_\lambda$-plane. The question of whether dV/dt (where V is the volume of the universe) changes sign near the singularity or not is more complicated. If we return to the Ω-time formulation of the Kantowski-Sachs problem, we can carry over the analysis of the Bianchi type I case readily to give $d\Omega/dt = -1/N_\Omega$ and have $N_\Omega = \dfrac{3\sqrt{2}\,R_0^3}{16\pi^{5/2}}\,e^{-3\Omega}\,H_\Omega^{-1}$. ($A_\Omega$ means the quantity A in the Ω-time formulation). In order to give the behavior of N_Ω in terms of our solution it is necessary to know H as a function of λ-time variables. Using our transformation laws, we find $H^2 = p^2 - 3R_0^4\, e^{2\,-4\Omega} = \dfrac{4}{3}(p_{\beta_\lambda} - \dfrac{1}{2}p_{\Omega_\lambda})^2 - 3R_0^4\, e^{-2\sqrt{3}\Omega_\lambda}$. From our solution we see that the one zero of H is at $p_{\Omega_\lambda} = 0$. As we approach the singularity $p_{\Omega_\lambda} \to p_{\beta_\lambda} \to const$, we see that H does not go to zero in this region. The quantity N does go to zero as $\Omega \to \infty$ but only reaches it at $\Omega_\lambda = \Omega = +\infty$, in the same manner as in the Friedmann universes.

Classically, then, there is no hope of escaping the singularity. We must turn to the quantum-mechanical solution.

b. Quantum-Mechanical Singularity

Near the two singularities we are in the asymptotic region far from the "potential", in which the wave function is $g_\pm(k)e^{i\sqrt{3}k(\pm\Omega_\lambda-\beta_\lambda)}$, where $g_\pm(k)$ are well determined functions of k. If we make wave packets of these functions, they should move smoothly toward $\Omega_\lambda = +\infty$ as the wave packets did in the Bianchi type I case. Because there seems to be no help in avoiding the singularity by motion of wave packets we must, as we did for Bianchi type I universes, look at the behavior of the operator N^{-1}. We can turn H^2 of the previous subsection into a Hermitian operator and investigate what it does to our asymptotic wave functions. If we use the same criteria we did in the Bianchi case, it is not difficult to see that N does not pass through the zero operator at any time, therefore, this universe does not avoid a singularity even quantum-mechanically.

It is not difficult to see in this case that the reason the Kantowski-Sachs universe is so well-behaved near the singularity is that this region is far from the "potential" which makes the motion qualitatively different from that of Bianchi-type I universes. One is reminded of computer simulations[55] of non-relativistic quantum mechanics in which a sedate, well-behaved wave packet approaches an impenetrable barrier. When the packet "strikes" the barrier, the wave packet changes its shape drastically and continues to do so until it moves away from the barrier when it again becomes a smooth, well-behaved packet. In our case the "barrier" is the "potential" which causes the universe to turn around near the present epoch. Thus it seems that the universe is closer to displaying the more drastic effects of quantum mechanics now than it was or will be near the two singularities. We shall see examples of this curious behavior in other models described below.

3. Friedmann-Robertson-Walker Universes

a. Classical Singularity

Very little needs to be said about the classical singularity of Friedmann-Robertson-Walker universes. It occurs at $\Omega = +\infty$ and because of the fact that $\rho \propto \mu e^{3\Omega}$, this singularity, as is well known, is an infinite-density singularity. Because of the fact that this singularity is so well known, we shall say no more about it.

b. Quantum-Mechanical Singularity

In the ADM solution of Nutku[27] there is no dynamical behavior of the wave function. It is purely a standing wave. Whether or not there is a singularity depends on whether $\psi(\Omega = +\infty)$ is zero or not. As we have seen, the Nutku solution has $\psi(\Omega = +\infty) \neq 0$, while the DeWitt[6] solution has $\psi(\Omega = +\infty) = 0$ by the introduction of a hard wall at $\Omega = +\infty$. Whether one believes that quantum mechanics predicts no singularity or not depends on which solution one accepts. We shall discuss this further in a later section.

In the Friedmann-Robertson-Walker universes the question of the sign of the operator N^{-1} is simplest. Because $H^2 = 36\pi^2 R_o^2 k e^{4\Omega} k e^{-4\Omega} + 384\pi^2 \mu e^{-3\Omega} R_o^3$ for dust, say, this operator does not go through zero until $\Omega = \infty$, so H can never change sign until it is too late.

4. *Bianchi Type IX Universes*

a. The Diagonal Case, Classical and Quantum

Classically, Bianchi type IX universes seem to be
qualitatively different from Bianchi type I universes, in that regions
of the $\beta_+\beta_-$-plane are closed off from the universe point by the potential.
Near the singularity, however, the walls associated with the triangular
potential have moved out to infinity and the universe spends most of
its time moving under a Hamiltonian which closely resembles that of
Bianchi type I universes. The changes in direction of the universe
point as it bounces from the walls do not seem to imply any drastic
change in the underlying structure of the universe. In fact, the only
way in which bounces from the walls could change the behavior significantly from the type I case would be to allow H to go through zero at
$\Omega < \infty$. Of course, H does go through such a zero when the curvature
term $(\sim -e^{-4\Omega})$ takes over near the present epoch and causes the universe
to turn around. Near the singularity, however, this term is negligible,
and bounces from the wall (see Appendix C) only allow the empty
universe Hamiltonian to decay as $1/\Omega$ rather than remain a constant. Thus,
there is no way that H can go through zero until $\Omega = +\infty$, when it is too
late.

Classically, then, the diagonal type IX universe cannot escape a
singularity. We shall see if quantum mechanics is any help.

We could, using the wave functions of Section IIIE, construct a wave packet and follow it around to see that as we approach the singularity it is well-behaved. Misner[21], however, uses the energy eigenfunctions of a triangular well (which are also discussed in Section IIIE) to indicate that if the universe is classical now it remains classical back to the singularity. He does this by pointing out that the energy eigenvalues should be approximated by

$$E_n \sim (\tfrac{2}{3} \pi)^{3/4} |n| \Omega^{-1}, \tag{5.4}$$

where $|n|$ is some combination of the quantum numbers of a triangular box. The classical adiabatic invariant $H\Omega$ can be used to show that for large n, $E_n \Omega$ is adiabatically invariant or that

$$<n> = const., \tag{5.5}$$

where $< >$ means "average".

Misner[21] argues that the fact that n is adiabatically constant implies that if $|n|$ is large now (the universe behaves classically) then it must have been large in the past, or that the universe should have shown no quantum effects near the singularity and been unable to avoid the singularity. Zapolsky[13] has speculated that because the universe would be represented by a wave packet made up of a large number of eigenstates of the Hamiltonian, this argument may not be valid. If we look at the wave-packet picture in Section IIIE, we see that there is no indication of suspicious behavior, but that we cannot easily use our argument of H^2 going to zero as a criterion for possible change of sign in H. We find that H^2 changes discontinuously at each bounce, so if H^2 is small enough (near the singularity) H may change signs during one of these jumps. Without an exact solution we cannot be sure. Indications from other solutions, such as the Kantowski-Sachs case seem to say that this would not happen, but that H^2 would decay smoothly, much as in the classical case. Such an argument is, of course, persuasive rather than rigorous.

b. The Symmetric and General Cases, Classical and Quantum

If we reason by means of the Raychaudhuri equation, the symmetric case would be the first in which we would expect that there would be a possibility of avoiding a singularity. This is the first of the cases which has non-zero rotation. Of course, the singularity theorems tell us that we cannot avoid a singularity even in this case. We might expect, however, that there might be a major change in the character of the singularity.

If we examine the behavior of the universe point in the $\beta_+\beta_-$-plane, we see that no such major change occurs. The rotation wall affects the motion of the universe point very little and while the centrifugal wall moves away from the β_+-axis, as was pointed out, the area closed off from the universe point becomes smaller and smaller relative to the area of the total quadrangle formed by the walls.

The only possible true difference in the singularity could be caused by the behavior of ϕ. We see that ϕ changes by $\Delta\phi$ every time the universe point bounces off the centrifugal wall. There is a zero of $\Delta\phi$ which occurs at the singularity which in a 3-dimensional representation[12] would imply an instantaneous reversal of direction by the universe point. The natural question

is whether this behavior is real or a coordinate effect. From Eq.(5.2)
we see that $\Delta\phi$ is dependent only on the parameter $a = \sqrt{3}\mu C/H\cos(\zeta_{in})$,
so $H \rightarrow 0$ has the same effect on a that C becoming infinite would, that
is, the effect is as real as it would be if the rotation tensor[56] became
infinite. The question which remains, then, is whether infinite rotation
can be removed by coordinate changes.

In order to attempt such a coordinate change we restrict ourselves to a single orbit in the $\beta_+ = 0$ plane. In Section VI we show
that when V and V_r are neglected in H, Hamilton's equations for ϕ are
equivalent to a geodesic equation on a two dimensional hyperboloid
imbedded in a flat, Lorentz three-space.

In fact, if we write the equations of the hyperboloid as
$t = (1/2\sqrt{3})\cosh(2\sqrt{3}\beta_-)$, $x = (1/2\sqrt{3})\sinh(2\sqrt{3}\beta_-)\sin(2\phi)$,
$y = (1/2\sqrt{3})\sinh(2\sqrt{3}\beta_-)\cos(2\phi)$, where x, y and t are the coordinates in
the flat three space, the entire family of possible geodesics for
which $\phi = \frac{\pi}{4}$ at $\beta = \beta_{min}$ are generated by Lorentz transformations that
leave y invariant and have $\gamma = \sqrt{1 + a^2}$, operating on the generic geodesic
$x = 0$, $y = \pm \sinh(2\sqrt{3}\beta_-)$, $t = \cosh(2\sqrt{3}\beta_-)$. This geodesic is a line
through $\beta_- = 0$ along the line $\phi = 0$, $\phi = \frac{\pi}{2}$ in the $\beta_+ = 0$ plane. Thus
we can, by Lorentz transformations, bring any geodesic into this
generic form. If upon such transformation, V is changed very little,
then an observer traveling with the universe point will see very little
measurable difference between motion along the generic geodesic and
motion along any other. If we perform such a transformation on $V(\beta_-)$
in the $\beta_+ = 0$ plane, we find that

$$V(\beta_-) = (4/3)(cosh^2(2\sqrt{3}\beta_-) - cosh(2\sqrt{3}\beta_-)) \qquad (5.6)$$

becomes

$$V(\beta'_-, \phi') = (4/3)(\gamma^2[cosh(2\sqrt{3}\beta'_-) - \oint sinh(2\sqrt{3}\beta'_-)sin(2\phi')]$$

$$- \gamma[cosh(2\sqrt{3}\beta'_-) - \oint sinh(2\sqrt{3}\beta'_-)sin(2\phi')]) \qquad (5.7)$$

where $\oint = 1 - (1/\gamma^2) = \dfrac{a^2}{(1+a^2)}$ which goes to one as $a \to \infty$.

Figure (5.3) shows the wall associated with $V(\beta_-, \phi)$ and the wall of $V(\beta'_-, \phi')$ after transformation. While the potential changes greatly, the change does not bring the wall into the center of the potential, and if we say that V affects the behavior of the universe point only during collision with the wall, then an observer should not be able to detect a change in V upon transformation. Since V measures the curvature of the universe, it would take very subtle measurements of the curvature of space to tell an observer whether he was in a rotating or a non-rotating universe near the singularity.

The Singularity. As we have mentioned, even though the work of Hawking, Penrose, and Ellis[31] showed that rotation cannot eliminate the singularity, it has been conjectured that, because of the way it enters into the Raychaudhuri equation, rotation, once its effects were understood, would be shown to have a great effect on the *nature* of the singularity. Another argument based on the assumption that rotation energy would behave like some sort of matter energy says its effect would tend to wash out as the effects of matter do as the universe approaches the singularity. Both of these ideas turn out to be simultaneously

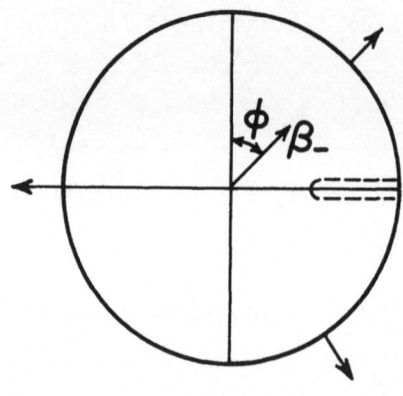

Before

Figure 5.3a. The potential $V(\beta_+, \beta_-, \phi)$ shown in section at constant β_+ before a Lorentz transformation in (x, y, t)-space is performed.

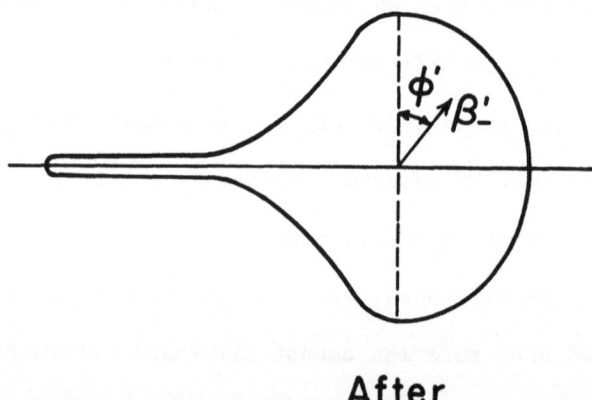

After

Figure 5.3b. The same potential after this Lorentz transformation has been performed.

right and wrong. If we examine the Hamiltonian, we see that rotation

appears in two guises. In the matter term it appears as the rotation

wall, and as the universe point cannot collide with the rotation wall,

there can be no qualitative change in the behavior of the universe

from this source. Thus, this effect of rotation washes out in the

manner to be expected if it were some type of energy.

In the centrifugal term rotation appears more subtly. While we

might expect, as in other models, that curvature (geometry) would

dominate the behavior of the universe near the singularity, the space

constraints (which are the R_{oi} Einstein equations) tie what might be

called rotation of the geometry (R_{oi}) to the constants describing the

rotation of matter. Thus, when matter has faded to a wisp, this rotation

of the geometry remains and has the possibility of causing large effects.

While the centrifugal potential can cause some large-scale effects,

we can see that it should have no effect on the singularity. In the

diagonal case the singularity occurs when $\Omega \to \infty$, where the triangular

potential V has become very large, leaving the universe point to

move in a straight line across the $\beta_+\beta_-$-plane. In order to compare

the behavior of the universe point in the diagonal and symmetric cases,

it is necessary to note that we can represent the diagonal case as a

universe point bouncing in a potential of the form ABCD in Fig. (3.6)

(with $\beta_c^{wall} = 0$ and $\beta_r^{wall} = \infty$) and ϕ being a two-valued function whose

value alternates between 0 and $\frac{\pi}{2}$ at each bounce from the β_+-axis. With

this we note that the symmetric and diagonal cases are similar, but

in the symmetric case a portion of the $\beta_+\beta_-$-plane is closed off by the

centrifugal potential and $\Delta\phi \neq \pi/2$ at each bounce. We have shown, however, that by a Lorentz transformation in superspace $\Delta\phi$ can be made to be $\pi/2$ at each bounce, and in relation to the triangular part of the potential the part of the plane closed off by the centrifugal potential becomes smaller and smaller. Thus the universe point ends up, as it does in the diagonal case, ranging in straight lines over wider and wider regions of the $\beta_+\beta_-$-plane. In this sense the singularity of the symmetric case is extremely close in character to the singularity of the diagonal case.

Because the quantum-mechanical symmetric case has not been fully worked out we shall discuss it only briefly. In Section IIIE we showed that if we treat μ and C which appear in the Hamiltonian H as c-numbers, we can quantize the model without difficulty following the full ADM prescription. If we treat H in this way the quantum problem is similar to that of the diagonal case. Because of the way we handled the problem in that case, we can readily see that the results carry directly to this case. The only difference between the two problems is the existence of ϕ. In the quantum-mechanical symmetric case we quantize after we have restricted ourselves to $p_\phi = \mu C$, so ϕ should be a straightforward operator constructed from the solution to the $\dot{\phi}$ equation. It seems to have little possibility of changing the singularity drastically.

The General Case. **Because** of the symmetry of the potentials in the general case we can see that the results we obtained for the symmetric case above can be used to show that the singularity of a general type IX universe

has the same general character as that of the diagonal case.

The fact that the rotation walls open and close does not obviate

the fact that this motion is superimposed on a general velocity of

magnitude one which keeps the universe point from ever catching any

of these walls. This implies that these walls, as in the symmetric

case have little effect on the nature of the singularity.

The problem of the centrifugal potentials is, as it is in the

symmetric case, more complicated. Because of the triangular symmetry

of both the gravitation potential and the three centrifugal potentials,

if one approaches one of the centrifugal potentials one can, by relabl-

ing the axes reduce the problem to the same problem as in the symmetric

case, with the added complication that the centrifugal potential and

the equation for the change of ϕ contain an opacity, a function of

γ and λ. If the opacity remains constant (this is possible because

while the regions near the centrifugal walls are tumbling regions it

is perfectly possible for the changes in γ and λ to leave the wall

we are considering unchanged) during reflection from the centrifugal

wall, the problem reduces to the same problem as in the symmetric case.

The case where this opacity changes during a bounce is more complicated.

Because, however, in the symmetric case one Lorentz transformation

caused no noticeable change in observations, a continuous series of such

transformations such as would be required if γ and λ changed should

cause no more effect than one. Therefore, we can, as we did then, remove

the effects of rotation when the universe point bounces from one

centrifugal wall. There still remains a region around $\beta_+ = \beta_- = 0$

where more than one centrifugal potential acts. In this region we still should be able to remove the effect of these walls with Lorentz transformations in higher dimensions.

The above discussion, we argue, is sufficient to show that the singularities of general type IX universes are much the same as those of symmetric type IX universes and hence close to those of the diagonal case.

Since the classical solution to the general case is in a less well defined state than the classical solution for the symmetric case, to attempt a quantum-mechanical solution is even more dubious than in the symmetric case. We *can* say, however, that unless some very subtle effect were to come in, the quantum-mechanical solution, because of the general form of the potentials and the way they enter in the classical case, should be similar enough to that of the symmetric case that we should not be able to avoid a singularity.

C. The Homogenization of Matter

Another use to which Hamiltonian cosmology may be put is the study of the "mixmaster universe" of Misner. Misner[24] has suggested that a type IX universe could be used to explain the remarkable isotropy of the 3°K black-body radiation. His thesis is that any given initial distribution of matter would tend to be homogenized as the type IX universe became less and less isotropic. Eventually the universe would reach a Friedmann-Robertson-Walker universe ($\beta_+ = \beta_- = 0$) with a homogeneous distribution of matter and radiation, in time to produce the isotropy of the black-body background. An essential component of this idea is the non-existence of horizons. In order for any process (such as, perhaps, shock waves) to smooth out the matter distribution it would be necessary for disturbances to be able to propagate throughout the universe. In the Friedmann-Robertson-Walker universes, for instance, this is impossible because of the existence of horizons; null geodesics cannot propagate far enough from any one particle before the time when the black-body radiation decouples to causally connect it to more than a fraction of the other particles in the universe. Misner[24] has suggested that because in some regions of the $\beta_+ \beta_-$-plane the universe is in some sense smaller in some directions than in others, geodesics could propagate along these directions and causally connect large sections of the universe. If motion in the $\beta_+ \beta_-$-plane is ergodic, then, this could occur in all directions in succession and eventually allow causal connection and, in consequence, homogenization of all the matter in the universe.

Chitre[9] has succeeded in showing that in type IX universes this
speculation is at least partly true, in that horizons are indeed
washed out by the ability of null geodesics to propagate entirely
around the universe if it remains in certain areas of the $\beta_+\beta_-$-plane
for long enough times. These areas are not a set of measure zero,
so if the motion of the universe point is ergodic, the washing-out of
horizons will indeed occur. We shall consider his proof of the existence
of these areas in four steps: 1) We shall give the geodesic equation
in type IX universes; 2) We shall study the Einstein equations for
type IX universes in a small region of the $\beta_+\beta_-$-plane; 3) We shall
study the Einstein equations in a larger region of the $\beta_+\beta_-$-plane;
4) We shall connect the geodesic equations and the Einstein equations
to show the non-existence of horizons.

1. The Geodesic Equations.

We shall consider null geodesics along the $\tilde{\sigma}^3$ direction in
a type IX universe, that is, we consider the null vector $\vec{v} = v^\mu \vec{e}_\mu$,
where $\vec{e}_0 = \frac{\partial}{\partial t}$ and $\vec{e}_i$ are the vectors dual to the forms $\tilde{\sigma}^i$, in which
$v^1 = v^2 = 0$. It is easily shown that $\vec{e}_3 = \frac{\partial}{\partial \psi}$. It is actually more
convenient to go to the orthomonal frame, $\tilde{\omega}^0 = dt$, $\tilde{\omega}^i = -\sqrt{6\pi}e^{\Omega}e^{-\beta^{}{}^{ij}}\tilde{\sigma}_j$ $(36\pi^2 R_o^4 = 1)$.
In this frame $\vec{e}_3 = -\sqrt{6\pi}\ e^{\Omega}e^{2\beta_+} \frac{\partial}{\partial \psi}$. Chitre[9] solves the geodesic equation
for the v^μ and finds $v^3 = \pm v^0 \propto e^{\Omega}e^{2\beta_+}$. This means that
$\vec{v} = v^0 \vec{e}_0 + v^3 \vec{e}_3 = v^0 \frac{\partial}{\partial t} + v^3 (-\sqrt{6\pi}\ e^{\Omega}e^{2\beta_+}) \frac{\partial}{\partial \psi}$. This implies that
$\frac{d\psi}{dt} = -\sqrt{6\pi}\ e^{\Omega}e^{2\beta_+}$ or

$$\frac{d\psi}{d\Omega} = \frac{d\psi}{dt}\frac{dt}{d\Omega} = \frac{2}{H}\ e^{-2\Omega}e^{2\beta_+} \, , \qquad (5.8)$$

using $\frac{dt}{d\Omega} = -\frac{1}{N}$ from Section III E.

2. *The Einstein Equations for Small β_- and Small p_-.*

Chitre[9] considers the case in which the universe point moves toward the channel which surrounds the β_+-axis in the diagonal type IX potential. Asymptotically the potential for this case is $V(\beta) \sim 16\beta_-^2 \, e^{4\beta_+} + 1$ and we have

$$H^2 = p_+^2 + p_-^2 + 16\beta_-^2 \, e^{-4\Omega} e^{4\beta_+}. \tag{5.9}$$

Chitre makes the transformation $\beta_+ = \beta_0 + \Omega$ and is able to reduce Hamilton's equations to

$$\frac{d^2\beta_-}{d\beta_0^2} + \left(\frac{16 \, e^{4\beta_0}}{K^2}\right) \beta_- = 0, \tag{5.10}$$

where $K = [p_+^2 + 16\beta_-^2 \, e^{4\beta_0}]^{1/2} - p_+$ is a constant. The solution to this equation is

$$\beta_- = Z_0\left(\frac{2e^{2\beta_0}}{K}\right), \tag{5.11}$$

where Z_0 is a Bessel function of order zero.

3. *The Einstein Equations for Large β_-.*

We want to consider the case where $\beta_- > 1$ and the universe point is moving almost parallel to the β_+-axis in the positive β_+ direction. In this situation the universe point is chasing the wall of the triangular potential which is parallel to the line $3\beta_- - \sqrt{3} \, \beta_+ = 0$. Chitre defines β_0 by $\beta_0 = \beta_+ + \Omega$, and shows that $K = H - p_+$ is a constant. Because $\frac{d\beta_0}{d\Omega} = \frac{d\beta_+}{d\Omega} - 1$, and $\frac{d\beta_+}{d\Omega} = \frac{p_+}{H}$, we find

$$\frac{d\beta_0}{d\Omega} = -\frac{K}{H}. \tag{5.12}$$

Since the track of the universe point is almost parallel to the β_+-axis, the angle of incidence with the wall it is approaching is almost $\pi/3$. Let us call it $(\pi/3 - \delta\theta)$. Eventually, then, the universe point will collide with this wall. We want to compute $\Delta\beta_0$ during this collision. From Eq. (5.12) we find $\Delta\beta_0 \approx - \frac{K}{H_{mean}} \Delta\Omega$. We find that $K = H_{in}[1-\cos(\delta\theta)]$ and if we take the mean value of H, H_{mean}, during the collision to be H_c of Appendix C , and $\Delta\Omega$ from that same Appendix, then

$$\Delta\beta_0 \approx - \frac{3\sqrt{2}[1-\cos(\delta\theta)]}{5-4\cos(\pi/3-\delta\theta)} \sqrt{\frac{2-\cos(\pi/3-\delta\theta)}{2\cos(\pi/3-\delta\theta)-1}} \ . \tag{5.13}$$

Since $\delta\theta$ is supposed to be small, $\Delta\beta_0 \approx \frac{4\sqrt{3}}{2} (\delta\theta)^{3/2}$.

4. *The Removal of Horizons*

In order to show that there exist epochs when geodesics can circumnavigate the universe, we shall investigate geodesics in the $\frac{\partial}{\partial\psi}$ direction during those times when the universe is moving nearly parallel to the β_+-axis.

We shall first consider the case when β_- is small. Eq. (5.11) gives us

$$\beta_- = Z_0 \left(\frac{2e^{2\beta_0}}{K}\right). \tag{5.14}$$

From Eq. (5.8) we have $\frac{d\psi}{d\Omega} = \frac{2}{H} e^{-2\Omega} e^{2\beta_+} = \frac{2}{H} e^{2\beta_0}$. The change in ψ as we move from Ω_1 to Ω_2 is

$$\Delta\psi = \int_{\Omega_1}^{\Omega_2} d\psi = \int \frac{2}{H} e^{2\beta_0} \frac{d\Omega}{d\beta_0} d\beta_0, \tag{5.15}$$

or

$$\Delta\psi = -\frac{2}{K}\int_{\Omega_1}^{\Omega_2} e^{2\beta_0}d\beta_0 = \frac{1}{2}\Delta\left(\frac{2e^{2\beta_0}}{K}\right) \tag{5.16}$$

Since circumnavigation of the universe is achieved when $\Delta\psi = 4\pi$, we need a $\Delta\left(\frac{2e^{2\beta_0}}{K}\right)$ of 8π. For small K (and proper initial choice of β_-),

$$\beta_- \sim \cos\left(\frac{2e^{2\beta_0}}{K} - \frac{\pi}{4}\right), \tag{5.17}$$

so β_- need only go through four cycles for its argument to change by 8π.

We shall next consider the large β_- case. Again we have

$$\Delta\psi = \frac{2}{K}\int e^{2\beta_0}\, d\beta_0. \tag{5.18}$$

During free flight $\Delta\beta_0 = 0$, but during collisions at large angles of incidence, β_0 changes by $\Delta\beta_0$, and

$$\Delta\psi \approx \frac{1}{K}\left(e^{2\beta_0}\right)_{in} 1 - e^{2\Delta\beta_0}$$

$$\approx \frac{1}{K}\left(e^{2\beta_0}\right)_{in} \sqrt[4]{3}\,(\delta\theta)^{3/2}. \tag{5.19}$$

Now $K = H_{in}[1-\cos(\delta\theta)] \approx \dfrac{H_{in}(\delta\theta)^2}{2}$, so

$$\Delta\psi \approx 2\sqrt[4]{3}\left(\frac{e^{2\beta_0}}{H}\right)_{in}(\delta\theta)^{-1/2}. \tag{5.20}$$

We would like to have a more obvious expression for $\left(\dfrac{e^{2\beta_0}}{H}\right)_{in}$. If the universe point has just bounced off the wall that is perpendicular to the β_+-axis and is heading for the slanting wall opposite, then the constant value which H has as it moves across the open part of the potential is $\dfrac{\sqrt{3}}{3}H_c$, where H_c is defined in Appendix C . Since H_c is

H at the moment of collision we approximate it as being given by $H^2 = \frac{1}{3} e^{-4\Omega^*} e^{-8(\beta_+)_{wall}}$ where Ω^* is Ω at collision. We know that β_0 is approximately constant during the motion toward the far wall, so $H^2 \approx \frac{1}{3} e^{-4\Omega^*} e^{-8(\beta_0)}$ in

$$\Delta\psi \cong 2 \sqrt[4]{3} \left(\frac{e^{2\beta_0}}{\frac{1}{\sqrt{3}} e^{-4\Omega^*} e^{-4\beta_0}}\right)_{in} (\delta\theta)^{-1/2} \qquad (5.21)$$

$$\cong 2(3)^{3/4} (e^{6\beta_+})_{wall} (\delta\theta)^{-1/2} .$$

Therefore, for all solutions for which $(\delta\theta) \leq \frac{(e^{6\beta_+})_{wall}}{(2\pi)^2 (3)^{3/2}}$, there is causal communication around the universe in the $\partial/\partial\psi$ direction. As Ω becomes large, $(\beta_+)_{wall}$ is large and negative, so there are sectors around lines parallel to the β_+-axis for which horizons are washed out. These sectors shrink as $\Omega \to \infty$, but are never strictly zero.

This argument establishes the washing out of horizons in the $\frac{\partial}{\partial\psi}$ direction The triangular symmetry of the potential insures this same process for the other two directions.

As a final note, Chitre[57] has made preliminary calculations which seem to indicate that the centrifugal potential in the symmetric case will cause little trouble in this argument if we take β_- large enough. This says that rotation will probably change the general result of this subsection very little.

VI. SUPERSPACE

Superspace, the space of all three-geometries, was first put forth by Wheeler[34,58] as the arena in which the development of the three-geometries of general relativity takes place. Superspace has been defined as the space of all Riemannian metrics (*Riem* (M)) modulo all possible transformations which give the same geometry by means of a different metric. We shall denote it by S. Recently this concept has found much use in Hamiltonian cosmology for much the same reason that the Hamiltonian approach is so compatible with cosmology. That is, the subspaces of superspace occupied by the cosmologies we have been considering are finite-dimensional, and complex general ideas about superspace reduce to simple, well-known concepts on these subspaces. We are, in this case, able to beg many complicated questions such as whether superspace should be defined in such a way that it is a manifold or not. This enables us to use the superspace concept to great advantage.

One thread of research into the structure of superspace has involved attempts to put some sort of metric on it. The first metric for superspace was proposed by DeWitt[6] . He makes use of the fact that the Hamilton-Jacobi approach of Peres[59] who writes the Hamilton-Jacobi equation for the Hamiltonian (2.19), say, as

$$(g_{ik}g_{j\ell} + g_{i\ell}g_{jk} - g_{ij}g_{k\ell})(\frac{\delta S}{\delta g_{ij}})(\frac{\delta S}{\delta g_{k\ell}}) + g^3 R = 0, \quad (6.1)$$

where $S = S(^3G)$, and δ signifies a functional derivative, contains the expression $(g_{ik}g_{j\ell} + g_{i\ell}g_{jk} - g_{ij}g_{k\ell})$ where one might expect to find

g^{ab} in a normal Hamilton–Jacobi equation for geodesics in spacetime. DeWitt[6], by requiring that coordinate transformations be isometries for a superspace metric and that the metric be local in the sense that different space points do not contribute to the distance, shows that a valid covariant superspace metric is

$$G_{ijk\ell}(x,x') = \frac{\sqrt{g}}{2} \, (g_{i\ell}g_{jk} + g_{j\ell}g_{ik} - 2g_{ij}g_{\ell k})\delta(x,x'), \qquad (6.2)$$

and the distance between two nearby metrics is

$$ds^2 = \int d^3x \, d^3x' \; G_{ijk\ell}(x,x')dg^{ij}dg^{k\ell} \qquad (6.3)$$

$$= \int d^3x \; d\bar{s}^2 \, ,$$

where $\;d\bar{s}^2 = \sqrt{g}(g_{i\ell}g_{jm} + g_{j\ell}g_{im} - 2g_{ij}g_{\ell m})dg^{ij}dg^{\ell m} \, .$

Misner, in unpublished notes, has proposed redefining the superspace metric by retaining the requirement that the superspace metric come from a quadratic form like $d\bar{s}^2$, but changes the definition slightly. That is, he defines

$$ds^2 = \int d^3V \, ds^2, \qquad (6.4)$$

$$ds^2 = (g_{i\ell}g_{jm} + g_{j\ell}g_{im} - 2g_{ij}g_{\ell m})dg^{ij}dg^{\ell m}, \qquad (6.5)$$

where d^3V is the invariant volume element $\sqrt{g} \; d^3x$. This metric is more convenient in the homogeneous case and Misner[60] has suggested that one can obtain an invariant Laplace-Beltrami operator by adding an appropriate additional term to the equation $\Box\psi = \psi^{;A}{}_{;A}$ proportional to the curvature of superspace to make $\Box$ a conformally invariant operator.

Homogeneous Cosmologies

We shall now restrict ourselves to homogeneous cosmologies and the subspace of superspace occupied by them (which Misner calls *mini-superspace*) and study the metric of Misner as applied to them (note that much, if not all, of what we say is true for inhomogeneous cosmologies also). Most of what follows comes directly from unpublished notes of Misner[60].

Superspace is supposed to be isometric under the transformation $g \rightarrow A^i{}_k g^{k\ell} A^j{}_\ell$, where the $A^i{}_j$ are constants (independent of g), so ds is invariant under them. We shall find it useful to compute the Killing vectors of this metric which are infinitesimal generators of these transformations. That is

$$X^i{}_j = (\frac{\partial}{\partial A_i{}^j}) A^i{}_j = \delta^i{}_j \quad = \quad \frac{\partial(A^T g^{**} A)}{\partial A} \frac{\partial}{\partial g^{**}} \quad = 2g^{ik} \frac{\partial}{\partial g^{kj}} \quad . \tag{6.6}$$

One particular Killing vector out of the nine given above is $X \equiv X^k{}_k$ which satisfies

$$[X, X^\ell{}_m] = 0, \forall \; \ell, m, \tag{6.7}$$

where [,] denotes the Lie bracket operation. The differential form $dg^{\ell m}$ operating on X gives $2g^{\ell m}$, so $ds^2(X,X) = -24$, so X is a time-like Killing vector in our metric.

We would like to use the Killing vectors we have just computed to relate the Einstein equations to geodesic-like equations for this superspace metric. We need for this to write the metric in contravariant form. If $ds^2 = G_{ijk\ell} \, dg^{ij} \otimes dg^{k\ell}$, then

(6.8)

$$\left(\frac{\partial}{\partial s}\right)^2 = G^{ijk\ell} \frac{\partial}{\partial g^{ij}} \otimes \frac{\partial}{\partial g^{k\ell}},$$

where $G^{ijk\ell} G_{k\ell rs} = \delta_{rs}^{ij} \equiv \frac{1}{2}(\delta_r^i \delta_s^j + \delta_s^i \delta_r^j)$. The contravariant metric is $G^{ijk\ell} = \frac{1}{2}(g^{ik} g^{j\ell} + g^{jk} g^{i\ell} - g^{ij} g^{k\ell})$. Misner[60] has shown that the $(\partial/\partial s)^2$ can be written as

$$\left(\frac{\partial}{\partial s}\right)^2 = \frac{1}{4} \left[X^i_{\ j} \otimes X^j_{\ i} - \frac{1}{2} X \otimes X \right].$$

(6.9)

We can compare this with the Hamiltonian constraint

$$\pi^i_{\ j} \pi^j_{\ i} - \frac{1}{2} \pi^2 = g^3 R$$

(6.10)

in the following way. If we consider a Hamilton-Jacobi functional $W(g)$, then $\pi_{ij} = \frac{\partial W}{\partial g^{ij}}$. From our definition of $X^i_{\ j}$ then,

$$\pi^i_{\ j} = \frac{1}{2} X^i_{\ j}[W],$$

(6.11)

$$\pi = \frac{1}{2} X[W].$$

(6.12)

This means that our Hamilton-Jacobi equation can be written as

$$\left(\frac{\partial W}{\partial s}\right)^2 = g^3 R.$$

(6.13)

We can show that this is very similar to the geodesic equation in superspace by writing (in a condensed notation where $ij \rightarrow A$) our Hamilton-Jacobi equation as

$$W_{,A} \, W_{,B} \, G^{AB} \; = \; R \equiv g^3 R. \qquad (6.14)$$

We may define a Hamiltonian associated with this equation as

$$H = \frac{1}{2} \, (\pi_A \pi_B \, G^{AB} \, - \, R), \qquad (6.15)$$

with the constraint $H = 0$. Hamilton's equations for this Hamiltonian are

$$\frac{dg^A}{d\lambda} = \frac{\partial H}{\partial \pi_A} \; = \; G^{AB} \pi_B \quad , \qquad (6.16)$$

$$\frac{d\pi_A}{d\lambda} = - \, \frac{\partial H}{\partial g^A} \; = \; - \frac{1}{2} \, \pi_C \pi_B \, G^{CB}{}_{,A} + \frac{1}{2} \, R_{,A} \qquad (6.17)$$

They can be combined to yield

$$\frac{d^2 g^A}{d\lambda^2} + \Gamma^A{}_{BC} \, \frac{dg^A}{d\lambda} \, \frac{dg^B}{d\lambda} \; = \frac{1}{2} \, R^{;A} \quad , \qquad (6.18)$$

or

$$v^A{}_{;B} \, v^B \; = \frac{1}{2} \, R^{;A} \quad , \qquad (6.19)$$

where $v^A \equiv dg^A/d\lambda$. This is obviously the geodesic equation in mini-superspace with a forcing term $\frac{1}{2} R^{;A}$ on the right-hand side. The constraint $H = 0$ becomes a normalization condition on v, that is it implies $v^A v_A = R$. This shows the connection between the Einstein equations and geodesics in mini-superspace with our definition of the metric.

Having arrived at Eq. (6.19), it is perhaps best to discuss another possible metric on superspace and its application to cosmology. DeWitt[61] has proposed a metric which differs from that of Eq. (6.2) by a conformal factor. This metric would take the form

$$\overline{G}_{ijk\ell} = [\int g^{\frac{1}{2}} {}^3R \; exp(-{}^3R_{ij} {}^3R^{ij}) d^3x]^{-1} \; exp(-{}^3R_{ij} {}^3R^{ij}) G_{ijk\ell}(x,x')$$

DeWitt[61] was able to show that such a metric leads to an equation of motion in superspace which is geodesic without a forcing term.

Gowdy[37] has extended this work and has applied it to diagonal Bianchi type IX universes. He writes the metric on the subspace of superspace inhabited by these universes as

$$ds^2 = (6\pi)^2 \, e^{-4\Omega}(1-V)(ds_M^2),$$

where ds_M^2 is the metric of Eq. (6.5). With this metric one can investigate the behavior of Bianchi type IX universes from a purely geometrical standpoint.

As a final note before we proceed to more concrete examples, we shall consider a method for finding constants of motion by means of the Killing vectors $X^i{}_j$. If we write

$$c = -\xi \cdot v,$$

where ξ is any of the Killing vectors $X^i{}_j$, then

$$\frac{dc}{d\lambda} = c_{,A}v^A = -(\xi_B v^B)_{,A}v^A$$

$$= -\xi_B v^B{}_{;A}v^A - \xi_{B;A}v^B v^A.$$

Killing's equation implies that the second term on the right is zero so

$$\frac{dc}{d\lambda} = -\frac{1}{2} R_{;A}\xi^A \ .$$

This implies that c will be a constant if the derivative of R along ξ vanishes, that is, if R is invariant under the symmetry generated by ξ.

Examples of the Metric on Mini-Superspace

We shall consider homogeneous cosmologies whose three-space metrics we can write in the form $g_{ij}dx^i dx^j = R_o^2 e^{-2\Omega}e^{2\beta}{}_{ij} \tilde\sigma^i\tilde\sigma^i$ where β_{ij} and Ω are independent of space coordinates and the $\tilde\sigma^i$ are space-dependent one-forms. In this case the parameters which determine the metric are Ω and the five independent β_{ij} and not the functions $S^i{}_j$ in $\tilde\sigma^i = S^i{}_j(x)dx^j$. This implies, because $g = S^T R_o^2 e^{-2\Omega}e^{2\beta}{}_{ij}S$, that $dg = S^T d(R_o^2 e^{-2\Omega}e^{2\beta}{}_{ij})S$ and that the metric (6.5) takes the form

$$d\Delta^2 = e^{-4\Omega}(e^{2\beta}{}_{ik}e^{2\beta}{}_{j\ell} + e^{2\beta}{}_{jk}e^{2\beta}{}_{i\ell} - 2e^{2\beta}{}_{ij}e^{2\beta}{}_{k\ell})d(e^{2\Omega}e^{-2\beta}{}_{ij})d(e^{2\Omega}e^{-2\beta}{}_{ij}).$$

Misner[60] has shown that this gives

$$d\Delta^2 = 24[-d\Omega^2 + \frac{1}{6} (d\beta)_{ij}(d\beta)_{ij}],$$

where $d\beta_{ij}$ is defined as usual. This metric has the

signature $- + + + + +$, as has been shown by DeWitt [6].

If we apply this metric to the Friedmann universes, we find
$ds^2 = -d\Omega^2$ and superspace is a one-dimensional Euclidean space. The

Kantowski-Sachs universe has

$$ds^2 = 24[-d\Omega^2 + d\beta^2],$$

so superspace is a flat, two-dimensional Lorentz space. Any diagonal

Bianchi type IX universe has

$$ds^2 = 24[-d\Omega^2 + d\beta_+^2 + d\beta_-^2],$$

a flat, three-dimensional Lorentz space. If we compare the Kantowski-

Sachs universe with the diagonal Bianchi type universes, we can see

the effect of the forcing term on the geodesic equation. From the

geodesic equation (6.19) and the fact that $v^A v_A = g^3 R$, we see that far

from the potential $(g^3 R \to 0)$ in the Kantowski-Sachs universe and always

$(g^3 R = 0)$ in the Bianchi type I universes, the universe point follows

null geodesics in Minkowski space, that is, straight lines. It is

only when $g^3 R$ is large (during collision with the potential) in the

Kantowski-Sachs case that we change from one null geodesic to another.

Note that the form of the equations implies that far from the potential

in Bianchi type IX universes the motion is geodesic.

Mini-superspace for the more complicated Bianchi types is easily

computed. For symmetric type IX universes we have

$$ds^2 = 24[-d\Omega^2 + d\beta_+^2 + d\beta_-^2 + \frac{\sinh^2(2\sqrt{3}\beta_-)}{3} d\phi^2],$$

the metric of the Cartesian product of a two-dimensional flat space and a two-dimensional space of constant, negative curvature. The most general Bianchi-type universe has

$$ds^2 = 24[-d\Omega^2 + d\beta_+^2 + d\beta_-^2 + \frac{\sinh^2(2\sqrt{3}\beta_-)}{3} (\tilde{\sigma}^3)^2 +$$

$$\frac{\sinh^2(3\beta_+ + \sqrt{3}\beta_-)}{3} (\tilde{\sigma}^2)^2 +$$

$$\frac{\sinh^2(3\beta_+ - \sqrt{3}\beta^2)}{3} (\tilde{\sigma}^2)^2].$$

DeWitt[6] has shown that $^6R = const.$ for this space.

We can get another view of superspace for the general case if we define three new variables, $r_1 = 2\sqrt{3}\beta_-$, $r_2 = -3\beta_+ - \sqrt{3}\beta_-$, and $r_3 = 3\beta_+ - \sqrt{3}\beta_-$. This makes our space the plane $r_3 + r_2 + r_1 = 0$ in the space

$$ds^2 = 24[-d\Omega^2 + \frac{1}{9} dr_1^2 + \frac{\sinh^2(r_1)}{3}(\tilde{\sigma}^3)^2 + \frac{1}{9} dr_2^2 + \frac{2\sinh^2(r_2)}{3} (\tilde{\sigma}^1)^2$$

$$+ \frac{1}{9} dr_3^2 + \frac{\sinh^2(r_3)}{3} (\tilde{\sigma}^2)^2].$$

If the σ's were exact the positive terms would be the Cartesian product of three two-dimensional spaces of constant, negative curvature. As it is, this form only indicates the structure of this subspace of super-space.

We can look at these subspaces from another viewpoint. All three are symmetric spaces, as has been pointed out by DeWitt. If we let $M_o(x)$ denote a trivial one-dimensional symmetric space (Euclidean) on the variable x, then we can write the superspace for the diagonal case as $M_o(\Omega) \times M_o(\beta_+) \times M_o(\beta_-)$. In the symmetric case the part of the metric dependent on β_+ is Euclidean, but the β_- part is more complicated. If we write $e^{2\beta}$ as

$$e^{2\beta_+} \, e^{-\phi \kappa_3} \begin{bmatrix} e^{2\sqrt{3}\beta_-} & 0 & 0 \\ 0 & e^{-2\sqrt{3}\beta_-} & 0 \\ 0 & 0 & e^{-6\beta_+} \end{bmatrix} e^{\phi \kappa_3} \, , \qquad e^{\phi \kappa_3}$$

leaves the 33 component unchanged. Its action on the β_- part is that of a rotation in two dimensions, thus, by the argument of Section I this 2 x 2 corner of the $e^{2\beta}$ matrix, for appropriate values of β_- and ϕ, is any 2 x 2 matrix of determinant one, that is, it is a member of the Lie group SL(2). For any constant ϕ_o, the superspace metric is invariant under the transformation $\beta \to e^{-\phi_o \kappa_3} \beta \, e^{\phi_o \kappa_3}$, and since $e^{\phi_o \kappa_3}$ is a member of the group SO(2), the β_- part of the matrix is a member of the Lie group SL(2)/SO(2). According to the classification of Helgason [12], this is a symmetric space of type AI of rank one. Superspace for the symmetric case is then,

$$M_o(\Omega) \times M_o(\beta_+) \times SL(2)/SO(2) \ .$$

By a similar argument superspace for the general case is

$$M_o(\Omega) \times SL(3)/SO(3) \ ,$$

where $SL(3)/SO(3)$ is a symmetric space of type AI of rank two. This discussion makes the progression from the diagonal to the symmetric to the general case clearer.

VII. QUANTIZATION

This section is intended mainly as a review of the techniques of
quantization of the cosmological models which we have discussed in
various sections. It is appropriate at this point to discuss the wider
meaning of quantized cosmologies. It must be emphasized that when we
quantize these models we are doing just that, quantizing models. We
have no assurance that if we add inhomogeneous degrees of freedom, any
of the conclusions we have reached, especially about singularities,
will be valid. The question which quantized models leave unanswered
is: What occurs in the tubes of superspace surrounding the tracks
of the homogeneous cosmologies? The quantum mechanical problem could
take two directions. If we consider simple examples we could examine,
say, the square well and see what happens as we increase dimensions.
The important quantum mechanical properties of one, two and three-
dimensional square wells are quite similar. The adding of dimensions
in superspace could produce a situation like this, in which the homogeneous
cosmologies give all the important behavior of any cosmology. An example
of the possibility in which quantization of a model would be misleading is
given by an artificial universe whose time development is the same as
that of the diagonal type IX universe, but which we endow with a true
singularity at $\beta_+ = \infty$. If we restricted ourselves to $\beta_- = 0$ and afterwards
quantized the model a wave packet could move in from $\beta_+ = \infty$ along the
β_+-axis channel, collide with the wall perpendicular to the β_+-axis
$(\sim e^{-8\beta_+})$ then return to $\beta_+ = \infty$. If we then let β_- be non-zero, we see
from Section III that, even classically, the universe could not reach

the singularity. In which of these two directions will the analysis of inhomogeneous cosmologies go? We cannot know this without more detailed study of inhomogeneous models near the homogeneous ones. One such model, which promises to be very useful in such a study is that of Belinskii and Khalatnikov which is inhomogeneous, but in which each space point is independent of all other space points. The study of models can only go so far, however. For instance, to make verifiable statements about the influence of quantum mechanics on the singularity we must have a full quantum analogue of the singularity theorems of Hawking and Penrose[31].

With the above ideas about the limits of validity of the procedure of quantizing models, we can discuss the quantization of homogeneous cosmologies where $g_{ij} dx^i dx^j = e^{-2\Omega(t)} e^{2\beta(t)}{}_{ij} \tilde{\sigma}^i \tilde{\sigma}^j$. If we neglect any problems that are due solely to the complexity of the resulting Hamiltonian (such as the fact that H is explicitly time-dependent), the only theoretical problems we encounter in quantizing the ADM Hamiltonian are the problem of factor ordering as we pass from the classical to the quantum problem, coupled with the fact that we end up with square-root Hamiltonians.

We shall, in the light of the models we have considered, discuss the three methods of handling such square-root Hamiltonians. These are:

1) the square-root method

2) the Dirac method

3) the SKG method.

As we pointed out in Section III A, the square-root method, if we allow positive and negative values for the square-root Hamiltonian, seems to offer no new information compared to the more well known

SKG approach. Because of this we shall not discuss this method here.

The Dirac method has, as we have pointed out in various sections,

two major shortcomings. The first of these is physical rather than

mathematical. While we are able in Bianchi types I and IX to produce

a linearized Hamiltonian on a state space with two spinor indices

$(\psi = \begin{pmatrix} \psi_1 \\ \psi_2 \end{pmatrix})$, our present knowledge about the universe gives us no physical

interpretation of the two components. This problem could be ignored

while one searched for a way of attaching physical meaning to this

wave function if it were not for the second objection. This is that

the "potentials" we have been discussing are not actually potentials,

but play the role of space-and-time-dependent masses, that is, we are

dealing with the equivalent of the following Hamiltonian,

$$H = \sqrt{\frac{\partial^2}{\partial x^2} + \frac{\partial^2}{\partial y^2} + m^2(x,y,t)} \ .$$

While we can linearize such a Hamiltonian, m does not commute with

$\frac{\partial}{\partial x}, \frac{\partial}{\partial y},$ and $\frac{\partial}{\partial t}$. This means that any attempt to recover H^2 as the

Klein-Gordon equation will encounter great difficulty in eliminating

terms in the various derivatives of m. Because of the two reasons we

have given, the Dirac method seems not to be too helpful in quantum

cosmology.

SKG quantization seems, by default, to be the most logical approach

to the quantization of cosmology. The major features favoring its use

seem to be its simplicity and the fact that it is well known. It

might be argued that the superspace formalism which gives the Einstein

equations in terms of a Hamilton-Jacobi system that is second order

points the way toward a second-order (i.e. SKG) wave equation. This
is not necessarily so. The Hamilton-Jacobi formulation of relativistic
particle mechanics is second order (cf., for example, Corben and Stehle[62]),
and the fact that from this we can obtain the Dirac equations seem
to indicate no necessary connection between the order of the Hamilton-
Jacobi equation and that of the quantum-mechanical wave equation. If,
however, we ignore the question of the order of wave equation, the Hamilton-
Jacobi equation tends to confirm our choice of canonical variables
and of the form of the wave equation.

We must, of course, mention the defects of the Klein-Gordon equation.
First, because it is second order in time, knowing ψ at any time does
not uniquely give ψ at another time unless we know $d\psi/dt$. Second, we
have the well-known problem of the non-positive-definiteness of the
probability density. If we are willing to work around these problems, SKG
quantization seems to be the best method at present to use in quantum
cosmology.

The last question we shall consider is that of constraints. This
is merely an extension of the classical problem. In the full ADM
prescription the space constraints are all solved before the true
Hamiltonian is considered to be discovered. If we want, as we have in
some cases, to leave some constraints to be solved in conjunction with
Hamiltonian, the quantum-mechanical problem will be quite different.
We can use the homogeneous cosmologies as a probe to compare these two
prescriptions.

Classically these two approaches differ only semantically when
applied to homogenous cosmologies. Solution of the constraints "in

conjunction with" the solution of the Hamiltonian can hardly be distinguished from substitution of the constraints and solving the resulting Hamiltonian.

Quantum-mechanically the two approaches can differ radically. The most striking example of this is in the symmetric type IX universe. The ADM procedure still requires the solution of the constraints before quantization of the Hamiltonian. The other approach leaves the constraints to reduce the solutions of the general Hamiltonian to physically reasonable systems. In the symmetric type IX case, not only does this lead to different ways of handling the operator ϕ, the wave equations we obtain for β_+ and β_- are different in the two cases.

If we think of the difference in the two equations for closed, Friedmann-Robertson-Walker universes, that of Nutku[27] and that of DeWitt[6], caused by differences in factor-ordering, we see that the differences in the two types of approach to symmetric type IX universes are also caused by a species of factor-ordering. In all quantization of cosmology we must ask ourselves if we are allowed to perform any classical operation before we quantize the system, that is, whether such operations commute quantum-mechanically. When they do not commute, we are left with more than one choice for the wave equation for the quantized universe. Lacking any experimental evidence to choose between one factor ordering and another, we must retreat to philosophical considerations to decide, which, of course, leaves any procedure one picks open to dispute.

APPENDIX A
LAGRANGIAN COSMOLOGY

The Lagrangian formulation for Bianchi-type cosmologies was developed by Misner[7] for type I universes and has been extended by him to diagonal type IX universes. Matzner[48] has extended this formalism to type V universes, and Hawking[14] has made it general. Matzner, Shepley, and Warren[32] have applied this method to symmetric type IX universes and Matzner[15] has begun to explore more complicated type IX universes containing non-fluid matter by means of it.

The Lagrangian formulation is based on the fact that the quantity $(G_{oo}-3(\Omega')^2)e^{-3\Omega}-8\pi T_{oo}e^{-3\Omega}, (A'=dA/dt)$, for *any* Bianchi-type universe, when varied with respect to β_{ij} (Ω taken as a known function of time) is a Lagrangian for the space part of the Einstein equations. This has been shown by Hawking[14].

The Lagrangian one arrives at is

$$L = \tfrac{1}{2} tr[(\sigma)^2]e^{-3\Omega} + \tfrac{1}{2} e^{-3\Omega}(^3R) - 8\pi T_{oo}e^{-3\Omega} , \qquad (A1)$$

where $\sigma_{ij} = \dfrac{d\beta_{ij}}{dt}$. We can compute $tr[(\sigma)^2]$ from our expression for $d\beta$ in Section II.

If we insert $d\beta_{ij}$ from Section II L becomes

$$L = [3(\beta_+')^2 + 3(\beta_-')^2 + sinh^2(2\sqrt{3}\beta_-)(\omega^3)^2 + sinh^2(3\beta_+ + \sqrt{3}\beta_-)(\omega^1)^2$$

$$+ sinh^2(3\beta_+ - \sqrt{3}\beta_-)(\omega^2)^2] + \tfrac{1}{2} e^{-3\Omega}(^3R) - 8\pi T_{oo}e^{-3\Omega} , \quad (A2)$$

where ω^i is the i'th component of the usual expression for angular velocity of a rigid body with respect to the body axis (See Ref. [62] p. 143).

We now want to define a "Hamiltonian" h after Misner[7] by

$$h = (\Omega')^2 \{3[(\dot{\beta}_+)^2 + (\dot{\beta}_-)^2] + sinh^2(2\sqrt{3}\beta_-)(\overline{\omega}^3)^2 + sinh^2(3\beta_+ + \sqrt{3}\beta_-)(\overline{\omega}^1)^2$$

$$+ sinh^2(3\beta_+ - \sqrt{3}\beta_-)(\overline{\omega}^2)^2\} - \frac{1}{2} e^{-3\Omega}(^3R) + 8\pi T_{oo}^{\cdot} e^{-3\Omega} , \qquad (A3)$$

$(\dot{A} \equiv dA/d\Omega, \ \omega^i \equiv \tilde{\sigma}^i/dt, \ \overline{\omega}^i \equiv \tilde{\sigma}^i/d\Omega)$, and an "energy" as

$$\Lambda = (\Omega')^2[e^{-6\Omega}(3[(\dot{\beta}_+)^2 + (\dot{\beta}_-)^2] + sinh^2(2\sqrt{3}\beta_-)(\overline{\omega}^3)^2 + sinh^2(3\beta_+ + \sqrt{3}\beta_-)(\overline{\omega}^1)^2$$

$$+ sinh^2(3\beta_+ - \sqrt{3}\beta_-)(\overline{\omega}^2)^2)] - \frac{1}{2} e^{-6\Omega}(^3R) + 8\pi T_{oo} e^{-6\Omega} . \qquad (A4)$$

From our definition of Λ we find that $3(\Omega')^2 = \Lambda e^{6\Omega}$, so

$$\Lambda = \Lambda\{(\dot{\beta}_+)^2 + (\dot{\beta}_-)^2 + \frac{1}{3} sinh^2(2\sqrt{3}\beta_-)(\overline{\omega}^3)^2 + \frac{1}{3} sinh^2(3\beta_+ + \sqrt{3}\beta_-)(\overline{\omega}^1)^2$$

$$+ \frac{1}{3} sinh^2(3\beta_+ - \sqrt{3}\beta_-)(\overline{\omega}^2)^2\} - \frac{1}{2} e^{-6\Omega}(^3R) + 8\pi T_{oo} e^{-6\Omega}. \qquad (A5)$$

From the equation of motion for h we find that $\frac{d\Lambda}{d\Omega} = \frac{\partial\Lambda}{\partial\Omega}$.

It is not difficult to see that $\Lambda = H^2$ of the Hamiltonian formulation with momenta replaced by the solutions of $\dot{\beta}_\pm = \partial H/\partial p_\pm$, $\dot{\phi} = \partial H/\partial p_\phi$, $\dot{\psi} = \partial H/\partial p_\psi$, $\dot{\theta} = \partial H/\partial p_\theta$. This implies that the Euler-Lagrange equations for the Lagrangian (A2), coupled with $3(\Omega')^2 = \Lambda e^{6\Omega}$ have the same dynamical content as Hamilton's equations for the Hamiltonian of Eqs. (2.19) and (3.40). The only thing we lack in this formulation are the space constraints $C^i{}' = 0$. ADM point out that these constraints are merely the equations $R_{oi} = 8\pi T_{oi}$, so these equations supply the constraints. Ryan[12] has shown how these equations in the symmetric type IX case lead to a problem equivalent to that obtained in the Hamiltonian formulation.

Because the two approaches, the Lagrangian and Hamiltonian formulations give the same behavior of the universes, it is only philosophical considerations which allow us to choose between them. The Hamiltonian formulation is more general and the derivation of the equations is straightforward, while the Lagrangian approach gives less justification for its equations. The Lagrangian formulation does have the advantage of inserting $8\pi T_{oo}$ and $8\pi T_{oi}$ explicitly without the need of constructing an L_M.

This appendix is perhaps the best place to note the work of Ellis, MacCallum, and Stewart[63], who have attacked numerous problems in Bianchi-type universes by means of non-Lagrangian approaches which owe much of their momenclature to the Lagrangian approach. They have concentrated on different types of matter in these universes and on observational consequences of various initial conditions.

APPENDIX B

THE QUANTUM BEHAVIOR OF AN EXPANDING, ONE-DIMENSIONAL, SQUARE WELL

We want to consider the problem of a relativistic, massless particle $(H^2 = p^2)$ moving in a one-dimensional square well that is expanding with velocity v (see Fig. B1). Note that v must be less than one for the problem to be meaningful, and in fact we would want $v = \frac{1}{2}$ to model the problem of the trianglar well in Bianchi type IX universes. The problem of a one-dimensional, expanding, square well has been considered by Jacobs, Misner, and Zapolsky[41], and this appendix is based mainly on their work.

In order to study this problem we must consider solutions of the Klein-Gordon equation which go to zero at $x = \pm vt \mp L_o$, where $\pm L_o$ are the positions of the walls at $t = 0$. Jacobs, Misner, and Zapolsky[41] find that the following set of functions satisfy the Klein-Gordon equation, $- \partial^2\psi/\partial t^2 + \partial^2\psi/\partial x^2 = 0$, and are zero at $\pm v\bar{t}$, where $\bar{t} = t - \frac{L_o}{v}$.

$$\psi_n(x, t) = exp\{-i\omega_n \ell n\ [(1+v)(\bar{t}-x)]\} - exp\{-i\omega_n \ell n\ [(1-v)(\bar{t}+x)]\}, \qquad (B1)$$

where $\omega_n = n\pi/\ell n(\frac{1+v}{1-v})$, $n = 1,2,\cdots$. They show that in the limit $v \to 0$ one recovers the usual wave functions for a static square-well.

If we assume that these wave functions are orthogonal at $t = const.$ (in the sense of $\int_{-v\bar{t}}^{v\bar{t}} \psi_n^* \psi_m\ dx = 0$ $m \neq n$ for $\bar{t}$ a constant) then we can consider after Zapolsky[13] the quantities $\langle\psi_n|H|\psi_n\rangle$, where $H = -i\frac{\partial\psi}{\partial t}$ as giving H in the sense that $\langle\psi|H|\psi\rangle = \sum_n a_n\langle\psi_n|H|\psi_n\rangle$ if $\psi = \sum_n a_n\psi_n$. So $H = const.$ will give $\langle\psi|H|\psi\rangle = const.$ we must divide by the normalization $\langle\psi|\psi\rangle = \sum a_n\langle\psi_n|\psi_n\rangle$, so we define $H_{nn} = \langle\psi_n|H|\psi_n\rangle/\langle\psi|\psi\rangle$. In order to

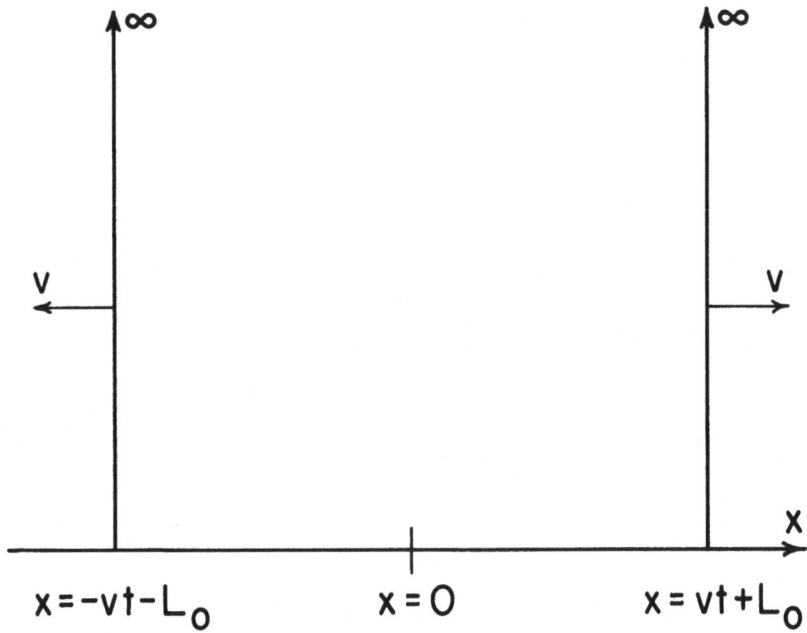

Figure BI. An expanding square well.

compute H_{nn} we need to know something about $\langle\psi_n|H|\psi_n\rangle$ and $\langle\psi_n|\psi_n\rangle$. We have

$$\psi_n^* H\psi_n = \frac{\omega n}{\overline{t}} \left[\left(\frac{1-v}{1-\eta}\right)\left(1-e^{i\omega_n \ell n \left(\frac{1+\eta}{1+\eta}\frac{1-v}{1+v}\right)}\right) + \left(\frac{1+v}{1+\eta}\right)\left(1-e^{-i\omega_n \ell n \left(\frac{1+\eta}{1-\eta}\frac{1-v}{1+v}\right)}\right)\right] \quad (B2)$$

where $\eta = x/\overline{t}$.

Since $\int f(x)dx = \overline{t}\int f(\eta)d\eta$ for $\overline{t} = const.$ we find

$$\langle\psi_n|H|\psi_n\rangle = 2n\pi - 2i\int_{-\ell n\left(\frac{1+v}{1-v}\right)}^{\ell n\left(\frac{1+v}{1-v}\right)} \tanh(y/2)\sin(\omega_n y)dy = C_n(v) \quad (B3)$$

We have

$$\langle\psi|\psi\rangle = \overline{t}\sum_n a_n^2 \left\{2v + \int_0^{2\ell n\left(\frac{1+v}{1-v}\right)} sech^2 y\, \cos(\omega_n y)dy\right\} = \overline{t}N. \quad (B4)$$

Note that $\langle\psi_n|H|\psi_n\rangle$ is not real, implying that H is not an Hermitean operator at $\overline{t} = const.$ If we define, after Zapolsky[13], an operator $H' = \frac{1}{2}(H+H^\dagger)$, then $\langle\psi_n|H'|\psi_n\rangle = 2\pi n$ and H' is Hermitean. If we consider H' to be the true "energy" operator then

$$H'_{nn} = \frac{2\pi n}{N\overline{t}} \quad (B5)$$

thus, if $\int \psi_n^* \psi_m\, dx\big|_{\overline{t}=const.} = 0$ then we can say that, quantum-mechanically, an expanding square-well has

$$E_n = \frac{A_n}{t}, \quad (B6)$$

where the A_n are constant.

In order to complete this appendix we must look at the quantities $\int \psi_m^* \psi_n\, dx,$ and $\int \psi_m^* H'\psi_n\, dx$ at $\overline{t} = const.$ We find

$$\langle\psi_m|\psi_n\rangle = \overline{t}\ exp\{\frac{i\ (m-n)\,\ell n\,\overline{t}}{\ell n\ (\frac{1+v}{1-v})}\}\int_{-v}^{v}[exp\{\frac{i\pi(m-n)\,\ell n\,[(1+v)(1-\eta)]}{\ell n\ (\frac{1+v}{1-v})}\}$$

$$+\ exp\{\frac{i\pi(m-n)\,\ell n\,[(1-v)(1+\eta)]}{\ell n\ (1+v\ /1-v)}\}$$

$$+\ exp\{\frac{i\pi\,\ell n\,([(1+v)(1-\eta)]^m/[(1-v)(1+\eta)]^n}{\ell n\ (1+v/1-v)}\}$$

$$+\ exp\{\frac{i\pi\,\ell n\,([(1-v)(1+\eta)]^m/[(1+v)(1-\eta)]^n}{\ell n\ (1+v/1-v)}\}]\ d\eta,$$

$$\equiv \overline{t}\ exp\{\frac{i\ (m-n)\,\ell n\,\overline{t}}{\ell n\ \{\frac{1+v}{1-v}\}}\}\ N_{mn}(v),\tag{B7}$$

and

$$\langle\psi_m|H'|\psi_n\rangle = -\ exp\{\frac{i\pi(m-n)\,\ell n\,\overline{t}}{\ell n(1+v/1-v)}\}\int_{-v}^{v}\left[\omega_m\left[\frac{exp(\frac{i\pi\,\ell n\,\{[(1-v)(1+\eta)]^m/[(1+v)(1-\eta)]^n\}}{\ell n\ (1+v/1-v)})}{(1-v)(1+\eta)}\right.\right.$$

$$+\ \left.\frac{exp(\frac{i\pi\,\ell n\,\{[(1+v)(1-\eta)]^m/[(1-v)(1+\eta)]^n\}}{\ell n\ (1+v\ /1-v)})}{(1+v)(1-\eta)}\right]$$

$$+\ \omega_n\left[\frac{exp(\frac{i\pi\,\ell n\,\{[(1+v)(1-\eta)]^m/[(1-v)(1+\eta)]^n\}}{\ell n\ (1+v/1-v)})}{(1-v)(1+\eta)}\right.$$

$$+\ \left.\left.\frac{exp(\frac{i\pi\,\ell n\,\{[(1-v)(1+\eta)]^m/[(1+v)(1-\eta)]^n\}}{\ell n\ (1+v/1-v)})}{(1+v)(1-\eta)}\right]\right]\ d\eta$$

$$\equiv exp\{\frac{i\pi(m-n)\,\ell n\,\overline{t}}{\ell n(1+v/1-v)}\}\ C_{mn}(v)\tag{B8}$$

Because of the complexity of the integrals in B7 and B8, we cannot be sure whether these cross terms are zero. In fact the form of the direct terms (B3,B4) seems to indicate that they are not. Basically, however, it is not important whether they are or not, because the functions $exp\{\frac{i\pi(m-n)\ell n \bar{t}}{\ell n(1+v/1-v)}\}$ are periodic in $\ell n\, \bar{t}$ with period $2\ell n\,(1+v/1-v)$. This means that they are periodic in $\bar{t}$ with expanding period, that is, the exponential is the same after a $\Delta \bar{t}$ of $\frac{4v}{(1-v)^2}\bar{t}$. Thus, if we look at $<H'>$ for $\psi = \sum_n a_n \psi_n$, we find

$$\frac{<\psi|H'|\psi>}{<\psi|\psi>} = \sum_n a_n^2 \frac{\pi n}{v\bar{t}} \, (1+f_n(\bar{t})) + \frac{1}{\bar{t}} \sum_{m \neq n} a_n a_m \, g_{mn}(\bar{t}) \qquad (B9)$$

where $f_n(\bar{t})$ and $g_{mn}(\bar{t})$ are periodic with period $\frac{4v}{(1-v)^2}\bar{t}$ and over long times have average values of zero. This means that $E\bar{t}$ is an adiabatic invariant equal to $\sum_n \frac{\pi n a_n^2}{v}$. Note from the form of the integrals that $f_n(\bar{t})$ and $g_{mn}(\bar{t}) \rightarrow 0$ as $v \rightarrow 0$, so we recover in that limit the usual energy levels for a square-well.

APPENDIX C

MISCELLANEOUS HAMILTONIAN CALCULATIONS

In this appendix we shall concentrate on the diagonal and symmetric type IX cases. These cases exhibit all the important features of the motion for any type IX universe except for reflections from centrifugal walls in tumbling regions, which are such complicated cases that we are forced to ignore them in the qualitative approach. The following calculations demonstrate all the relations which we have given in the body of the work without proof.

Reflections from Walls. During collisions of the universe point with the gravitation walls H changes, so we must investigate its behavior in order to determine the reflection law of the universe point. Since collisions with the gravitation walls will usually take place far from centrifugal walls we can take ϕ, θ, ψ constant during these encounters. In this case the problem is the same as that for the diagonal type IX universes investigated by Misner[21,24], and the relations to be developed are useful in studies of that problem.

Because of the triangular symmetry of the gravitation well, we need only consider collisions from the wall perpendicular to the β_+-axis, where the potential $V(\beta)$ is asymptotically $\frac{1}{3} e^{-8\beta_+}$, to obtain the behavior of H under any collisions with V. For collisions with this wall Jacobs, Misner, and Zapolsky[41] have shown that there exist two constants of motion, $p_- = H \dot{\beta}_-$ and $K = H(1 + \frac{1}{2} \dot{\beta}_+)$, K being the Hamiltonian of the system in a coordinate system moving with the wall. We can use these two constants to substitute for V in $\dot{H} = \partial H/\partial \Omega$ and arrive at an equation for H,

$$H(H)^{\cdot} = 6H^2 - 16KH + 2(p_-^2 + 4K^2). \tag{C1}$$

If we define $H_{out}^{in} = (4K \pm \sqrt{4K^2 - 3p_-^2})/3$, we can write the solution to our equation as

$$\Omega = \frac{1}{H_{in} - H_{out}} [H_{out} \, \ell n \, (H - H_{out}) - H_{in} \, \ell n \, (H_{in} - H)] \, . \tag{C2}$$

To characterize this solution completely, we need only point out what H_{in} and H_{out} are. Note that as $\Omega \to -\infty$, $H \to H_{in}$ and as $\Omega \to +\infty$, $H \to H_{out}$, so H is, as we would expect, constant before and after collision and H_{in} and H_{out} are these constants at these respective times. With this result we can define a law of reflection. We define θ_{in} and θ_{out} by means of Fig. (5.1).

Before and after collision we can write $\dot{\beta}_+ = \mp cos(\theta_{out}^{in})$ and $\dot{\beta}_- = sin(\theta_{out}^{in})$ respectively (note that $(\dot{\beta}_+)^2 + (\dot{\beta}_-)^2 = 1$). Using the constancy of p_- we find

$$\frac{H_{out}}{H_{in}} = \frac{sin(\theta_{in})}{sin(\theta_{out})} \, . \tag{C3}$$

If we use the constancy of K/p_-, we can solve for θ_{out} in terms of θ_{in} and obtain our law of reflection. This is

$$sin(\theta_{out}) = \frac{3sin(\theta_{in})}{5 - 4cos(\theta_{in})} \, . \tag{C4}$$

Note that since the wall velocity is $\frac{1}{2}$ the universe point will never collide with the wall unless $0 < \theta_{in} < \pi/3$ (it will eventually collide with *some* wall if $\theta_{in} \neq \pi/3$).

Reflection from the centrifugal wall in the symmetric case is specular. If we define ζ_{in} and ζ_{out} by means of Fig. (5.2), we see that H is a constant during the bounce because the centrifugal potential has no explicit Ω-dependence. Also p_+ is a constant of motion because V_c is independent of β_+. These two constants imply that $|p_+|$ is the same before and after collision, which implies that $\zeta_{in} = \zeta_{out}$.

Changes of H during Bounces. It is not difficult to see from Eqs. (C3) and (C4) that for collisions with the gravitation well

$$\frac{H_{out}}{H_{in}} = \frac{5 - 4\cos(\theta_{in})}{3} . \tag{C5}$$

As was mentioned above, H does not change during bounces from the centrifugal wall.

Time Spent in Collision. We define $\Delta\Omega$, the *collision time*, the time spent in collision for each angle of incidence θ_{in} as $\Delta\Omega = - \left(\frac{d\Omega}{dH}\right)_{min} \Delta H$, where $(d\Omega/dH)_{min}$ is the maximum value of $d\Omega/dH$. From Eq. (C2) we have

$$\Delta H = H_{out} - H_{in} . \tag{C6}$$

The maximum of $d\Omega/dH$ is at H_c, where $d^2\Omega/dH^2\big|_{H_c} = 0$. This gives

$$H_c = \sqrt{H_{out} H_{in}} , \tag{C7}$$

from which

$$\left(\frac{d\Omega}{dH}\right)_{min} = \frac{H_{out}}{(H_{in}-H_{out})(H_c-H_{out})} +$$

$$\frac{H_{in}}{(H_{in}-H_{out})(H_{in}-H_c)} \tag{C8}$$

and

$$\Delta\Omega = \left(\frac{1 + \sqrt{H_{out}/H_{in}}}{1 - \sqrt{H_{out}/H_{in}}}\right) . \tag{C9}$$

As a function of θ_{in} we have

$$\Delta\Omega = \sqrt{2}\sqrt{\frac{2 - cos(\theta_{in})}{2cos(\theta_{in}) - 1}} . \tag{C10}$$

The Validity of the Wall Approximation. As an aside, we want to show that approximating the potentials by walls is valid. We do this by showing that if $\delta\Omega$ is the free time between two generic collisions that $<\delta\Omega/\Delta\Omega>$ is large, where $<\ >$ means an average over all collisions at random angles of incidence. We can use the geometry of an expanding triangle to compute $\delta\Omega$.

Consider the situation where we begin in on one wall of the triangle of side L a distance βL below the β_+-axis moving toward wall ③ (see Fig. (C1)) with an angle of incidence θ_{in}. The point moves toward ③ for a time Ω', collides with ③ then moves toward ② for a time Ω'' before colliding with it. Since there are two collisions during this time, the average free time is $\frac{1}{2}$ $(\Omega' + \Omega'')$.

From the geometry of Figure (C1) we see that

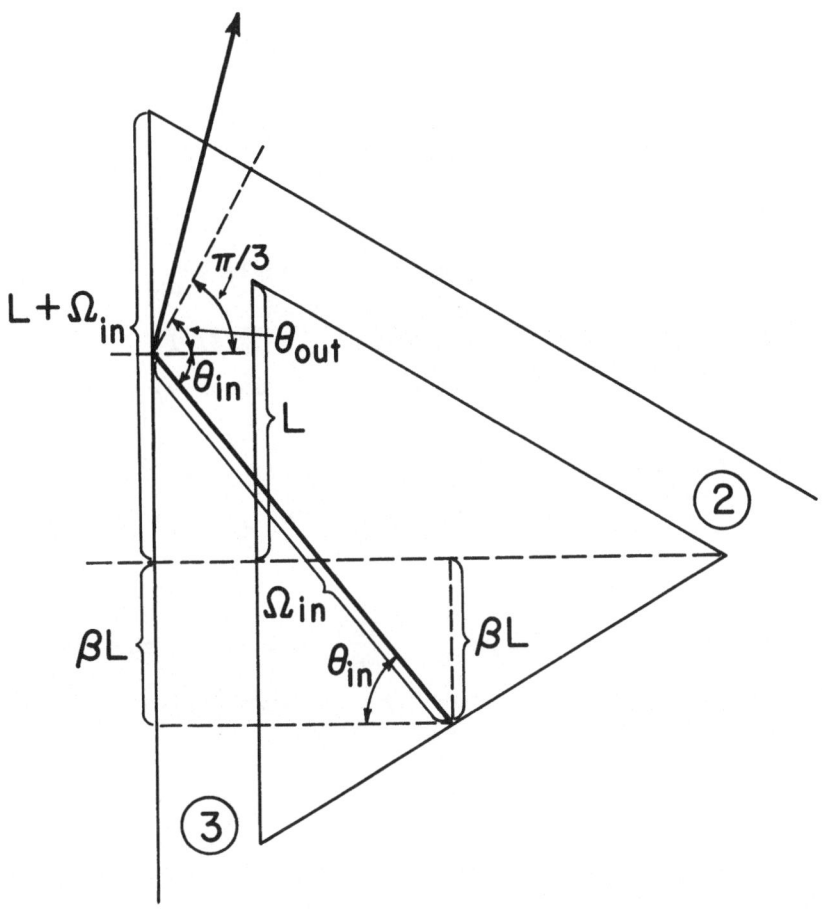

Figure CI. The geometry of an expanding triangle.

$$\Omega' = \frac{2\sqrt{3}(1-\beta)}{2\cos(\theta_{in}) - 1} \; . \tag{C11}$$

To find Ω'' first note that because the triangle, after expanding a time Ω', is still equilateral so, in Fig. (C1), $AB = \sqrt{3}[(1+\beta)L - \Omega'\sin(\theta_{in})]$. Now, since we are moving toward wall ② at an angle $(\theta_{out} - (\pi/3))$ with an initial distance AB we have $\Omega'' = \frac{2\,AB}{2\cos(\theta_{out} - \pi/3)-1}$. Inserting AB as above, and our expression for θ_{out} in terms of Ω_{in}, we find

$$\Omega'' = \frac{\sqrt{3}L([2\cos(\theta_{in})-1-2\sqrt{3}(1-\sin(\theta_{in}))]\beta+[(2\cos\theta_{in}-1)+2\sqrt{3}(1-\sin(\theta_{in}))](5-4\cos(\theta_{in}}{9(3\cos(\theta_{in}) + \sqrt{3}\sin(\theta_{in}) - 3)} \tag{C12}$$

Defining $L_1(\theta_{in}) = \frac{1\Omega'}{2\Omega}$, and $L_2(\theta_{in}) = \frac{1\Omega''}{2\Omega}$ and noting that because the walls expand with velocity one-half, $L = \Omega$ we arrive at

$$\delta\Omega = \Omega(L_1 + L_2) \quad . \tag{C13}$$

We can see from Eqs. (C11) and (C12) that $L_1 + L_2$ descends from ∞ at $\theta_{in} = 0$ to a minimum, then rises to ∞ again at $\theta_{in} = \pi/3$. By examining Eq. (C10) we can see that $\delta\Omega/\Delta\Omega$ has this same type of behavior. Equation (C13) shows us, however, that this minimum is porportional to Ω, so for large Ω, $\langle\delta\Omega/\Delta\Omega\rangle$ must be large, thus the universe point spends most of its time in regions where $H \approx constant$.

The Adiabatic Relation $H\Omega \approx constant$. Misner[8] has shown that for a special orbit in the $\beta_+\beta_-$-plane in the diagonal case that $H\Omega \approx constant$. We can show that in general this should be so. The most general motion of the universe point in the $\beta_+\beta_-$-plane is made up of a sum of segments like those traversed by the universe point in Figure (C1). Using the relations (C11) and (C12) we would like to examine the quantities $H_{in}\Omega'$ and $H_{out}\Omega''$, where H_{in} and H_{out} are the constant values of H before and after the collision respectively. If we define $\nu = H_{out}\Omega''/H_{in}\Omega'$ and take $\langle\beta\rangle = \frac{1}{2}$, we find

$$\nu(\theta_{in}) = \frac{\frac{1}{2}(3\cos(\theta_{in}) - \sqrt{3}\sin(\theta_{in}) - (\frac{3}{2} - \sqrt{3})(5-4\cos(\theta_{in}))^2}{27(3\cos(\theta_{in}) + \sqrt{3}\sin(\theta_{in}) - 3)} \,. \qquad \text{(C14)}$$

As a function of θ_{in}, ν descends from infinity at $\theta_{in} = 0$, crosses 1 at $\theta_{in} = 15.5°$ (the special case of Misner[8]), descends to a minimum less than one (for $\theta_{in} = 30°$, $\nu \sim \frac{1}{5}$) and rises again to infinity at $\theta_{in} = \pi/3$. The two infinities reflect the fact that for these particular angles the universe point spends a long time chasing one of the walls at an angle of incidence $\pi/3$. Because (as the corner-run solution of Section III E tends to indicate) something eventually causes the universe point to stop moving in these directions and return to bouncing in the triangle again, at these points ν is not truly infinite. Since these infinites are not strong ones (they appear only logarithmically $\langle\nu\rangle_{\theta_{in} \, ave}$) any cutoff will tend to make ν oscillate randomly above and below a value of one. Without being able to integrate $\nu(\theta_{in})$ we cannot say for sure that $\langle\nu\rangle = 1$. Thus we cannot be sure whether $H\Omega \approx const.$ or or whether it grows $(\langle\nu\rangle < 1)$ or decays $(\langle\nu\rangle > 1)$ as $\Omega \to \infty$. For the present we shall accept the special case of Misner[8] as typical and assume adiabatically that $H\Omega \approx const.$

REFERENCES

1. L. Rosenfeld, Ann. Physik 5, 113 (1930)
   Z. Physik 65, 589 (1930).

2. A list of references is given in the review article by P. Bergmann, Helvetia Phys. Acta Suppl. 4, 79 (1956).

3. A general field theory is given in: P. Dirac, Can. J. Math. 2, 129 (1950); Applications to general relativity are in:
   Proc. Roy. Soc. (London) A246, 326 (1958)
   Proc. Roy. Soc. (London) A246, 333 (1958)
   Phys. Rev. 114, 924 (1959).

4. See: F. Pirani and A. Schild, Phys. Rev. 79, 986 (1950).

5. A list and summary of the ADM papers are given in:
   R. Arnowitt, S. Deser, and C. Misner, in *Gravitation: An Introduction to Current Research*, edited by L. Witten (John Wiley & Sons, Inc., New York, 1962).

6. B. DeWitt, Phys. Rev. 160, 1113 (1967).

7. See: C. Misner, Astrophys. J. 151, 431 (1968),
   C. Misner, Phys. Rev. Letters 22, 1071 (1969).

8. See: C. Misner, in *Relativity*, edited by M. Carmeli, S. Fickler, and L. Witten (Plenum Press, New York, 1970).

9. D. Chitre, (to be published).

10. L. Hughston and K. Jacobs, Astrophys. J. 160, 147 (1970)
    K. Jacobs and L. Hughston (to be published).

11. Y. Nutku (to be published in Proceedings: Pittsburgh Conference on Relativity, July 13-17, 1970, University of Pittsburgh)
    Y. Nutku and B. Kobre (to be published).

12. M. Ryan, Ph. D. Thesis, University of Maryland (submitted to Ann. Phys. (N. Y.))

13. H. Zapolsky, unpublished notes for [41].

14. S. Hawking, Mon. Not. Roy. Astron. Soc. (London) 142, 129 (1969).

15. See: R. Matzner, Ann. Phys. (N. Y.) (to be published),
    R. Matzner, a series of three papers (to be published).

16. C. Misner, Phys. Rev. Letters 19, 533 (1967).

17. V. Belinskii and I. Khalatnikov, Zh. Eksp. Teor. Fiz. 57, 2163 (1969)
    (Sov. Phys. - JETP 30, 1174 (1970)
    Zh. Eksp. Teor. Fiz. 59, 314 (1970)
    (Sov. Phys. - JETP, to be published).

18. R. Kantowski and R. Sachs, J. Math. Phys. 7, 443 (1966).

19. L. Bianchi, Mem. Soc. It. della Sc. (dei XL), (3), 11, 267 (1897).

20. A. Friedmann, Z. Phys. 10, 377 (1922)
    Z. Phys. 21, 326 (1924)
    See H. Robertson, Rev. Mod. Phys 5, 62 (1933) for the basis of the
    name "Friedmann-Robertson-Walker universes".

21. C. Misner, Phys. Rev. 186, 1319 (1969).

22. L. Fishbone (to be published).

23. So named by K. Gödel, Proc. Intern. Congr. Math., 1950, Vol. I, pp. 175-181.

24. C. Misner, Phys. Rev. Letters 22, 1071 (1969).
    A non-Hamiltonian, non-Lagrangian discussion of this universe from
    the Einstein equations is given by V. Belinskii and I. Kalatnikov,
    Zh. Eksp. Teor. Fiz. 56, 1700 (1969); Sov. Phys. - JETP 29, 911 (1969).

25. I. Ozsváth (to be published).

26. K. Jacobs and L. Hughston (to be published).

27. Y. Nutku (to be published).

28. B. Schutz (to be published).

29. K. Kuchař (to be published).

30. Y. Nutku and B. Kobre (to be published).

31. For a review see: R. Partridge, Am. Scientist 57, 37 (1969);
    See also: S. Hawking and G. Ellis, Astrophys. J. 152, 25 (1968),
    R. Penrose, in Batelle Rencontres 1967, edited by
    C. DeWitt and J. Wheeler (W. A. Benjamin, Inc., New York,
    1968) Chap. 7.

32. R. Matzner, L. Shepley, and J. Warren, Ann. Phys. (N. Y.) 57, 401 (1970).

33. A. Doroshkevich and I. Novikov (to be published).

34. For a review see: J. Wheeler, Am. Scientist 56, 1 (1968).

35. C. Misner (to be published).

36. B. DeWitt in Ref. [6] and in *Relativity*, edited by M. Carmeli, S. Fickler, and L. Witten (Plenum Press, New York, 1970).

37. R. Gowdy (to be published).

38. See: Schweber, H. Bethe, and de Hoffmann, *Mesons and Fields* Vol. 1 (Row, Peterson & Co., Evanston, Ill., 1956) Chap. 2.

39. For the name see: E. Schrödinger, Ann. d. Physik *81*, 109 (1926)
W. Gordon, Z. Physik *40*, 117 (1926)
O. Klein, Z. Physik *41*, 407 (1927)
Note also: V. Fock, Z. Physik *38*, 242 (1926).

40. See: O. Heckmann and E. Schücking, in *Gravitation: An Introduction to Current Research,* edited by L. Witten (John Wiley & Sons, New York, 1968), and references therein;
K. Thorne, Astrophys. J. *148*, 51 (1967);
K. Jacobs, Astrophys. J. *153*, 661 (1968) and *155*, 379 (1969);
C. Misner, Astrophys. J. *151*, 431 (1968).

41. K. Jacobs, C. Misner, and H. Zapolsky (to be published).

42. L. Hughston and K. Jacobs, Astrophys. J. *160*, 147 (1970).

43. C. Misner and D. Sharp, Phys. Rev. *B136*, B571 (1964).

44. L. Hughston and L. Shepley, Astrophys. J. *160*, 333 (1970).

45. H. Zapolsky in Ref. [41],
V. Moncrief, Private communication
K. Jacobs, Private communication.

46. S. Schelkunoff, *Electromagnetic Waves* (Van Nostrand, N. Y., 1943) pp. 393-4.

47. C. Misner in Ref. [8].

48. R. Matzner, Astrophys. J. *157*, 1085 (1969).

49. A. Einstein and N. Rosen, J. Franklin Inst. *223*, 43 (1937).

50. A. Penzias and R. Wilson, Astrophys. J. *142*, 419 (1965).

51. R. Partridge and D. Wilkinson, Phys. Rev. Letters *18*, 557 (1967)
R. Partridge, Invited talk, Fourth Texas Symposium on Relativistic Astrophysics; Dallas, Texas (Dec. 1968).

52. A. Raychaudhuri, Phys. Rev. *98*, 1123 (1955).

53. We let $\theta = u^\mu{}_{;\mu}$ be the scalar of expansion and
$\Omega_{\mu\nu} = u_{[\mu;\nu]} + u_{[\mu;\alpha}u^\alpha u_{\nu]}$, $\Sigma_{\mu\nu} = u_{(\mu;\nu)} + u_{(\mu;\alpha}u^\alpha u_{\nu)} - \frac{1}{3}\theta(g_{\mu\nu} + u_\mu u_\nu)$
be the tensors of rotation and shear, respectively, after
J. Ehlers, Akad. Wiss. Lit. (Mainz) Abhandl. Math. Nat. Kl.
11 (1961).

54. See, for example, J. Wheeler in *Batelle Rencontres 1967*, edited by
C. DeWitt and J. Wheeler (W. A. Benjamin, Inc., New York, 1968).

55. See: J. Schwartz and E. Taylor, AFIPS Proceedings, Fall Joint
Computer Conf. *33* pt.2, 1285 (1968).

56. See Ref. [12] for the relation of C to $\Omega_{\mu\nu}$.

57. D. Chitre, Private communication.

58. See for example, J. Wheeler in Ref. [54] and in *Relativity, Groups
and Topology*, edited by C. DeWitt and B. DeWitt (Gordon and Breach,
New York, 1964).

59. A. Peres, Nuovo Cimento *26*, 53 (1962).

60. C. Misner, unpublished.

61. B. DeWitt in *Relativity*, edited by M. Carmeli, S. Fickler, and
L. Witten (Plenum Press, New York, 1970).

62. H. Corben and P. Stehle, *Classical Mechanics* (John Wiley & Sons, Inc.,
New York, 1960).

63. G. Ellis, J. Math. Phys. *8*, 1171 (1967)
J. Stewart and G. Ellis, J. Math. Phys. *9*, 1072 (1968)
G. Ellis and M. MacCallum, Commun. Math. Phys. *12*, 108 (1968)
M. MacCallum and G. Ellis (to be published)
M. MacCallum (to be published).

Lecture Notes in Physics

Selected Issues from
Lecture Notes in Mathematics

Beschaffenheit der Manuskripte

Die Manuskripte werden photomechanisch vervielfältigt; sie müssen daher in sauberer Schreibmaschinenschrift mit ausreichend großer Type geschrieben sein. Handschriftliche Formeln bitte nur mit schwarzer Tusche eintragen. Notwendige Korrekturen sind bei dem bereits geschriebenen Text entweder durch Überkleben des alten Textes vorzunehmen oder aber müssen die zu korrigierenden Stellen mit weißem Korrekturlack abgedeckt werden. Die reproduktionsfähigen Abbildungen (in Originalgröße) sollen in den Text eingeklebt werden. Falls das Manuskript oder Teile desselben neu geschrieben werden müssen, ist der Verlag bereit, dem Autor bei Erscheinen seines Bandes einen angemessenen Betrag zu zahlen. Die Autoren erhalten 50 Freiexemplare.

Zur Erreichung eines möglichst optimalen Reproduktionsergebnisses ist es erwünscht, daß bei der vorgesehenen Verkleinerung der Manuskripte der Text auf einer Seite in der Breite möglichst 18 cm und in der Höhe 26,5 cm nicht überschreitet. Entsprechende Satzspiegelvordrucke werden vom Verlag gern auf Anforderung zur Verfügung gestellt.

Manuskripte, in englischer, deutscher oder französischer Sprache abgefaßt, sind einzureichen bei: Springer-Verlag, 6900 Heidelberg, Postfach 1780.

Cette série a pour but de donner des informations rapides, de niveau élevé, sur des développements récents en physique, aussi bien dans la recherche que dans l'enseignement supérieur. On prévoit de publier.

1. des versions préliminaires de travaux originaux et de monographies

2. des cours spéciaux portant sur un domaine nouveau ou sur des aspects nouveaux de domaines classiques

3. des rapports de séminaires

4. des conférences faites lors de congrès ou de colloques

En outre il est prévu de publier dans cette série, si la demande le justifie, des rapports de séminaires et des cours multicopiés ailleurs mais déjà épuisés.

Dans l'intérêt d'une diffusion rapide, les contributions auront souvent un caractère provisoire; le cas échéant, les démonstrations ne seront données que dans les grandes lignes. Les travaux présentés pourront également paraître ailleurs. Une réserve suffisante d'exemplaires sera toujours disponible. En permettant aux personnes intéressées d'être informées plus rapidement, les éditeurs Springer espèrent, par cette série de «prépublications», rendre d'appréciables services aux instituts de physique. Les annonces dans les revues spécialisées, les inscriptions aux catalogues et les copyrights rendront plus facile aux bibliothèques la tâche de réunir une documentation complète.

Présentation des manuscrits

Les manuscrits, étant reproduits par procédé photomécanique, doivent être soigneusement dactylographiés type assez grand. Il est recommandé d'écrire à l'encre de Chine noire les formules non dactylographiées. Les corrections nécessaires doivent être effectuées soit par collage du nouveau texte sur l'ancien soit en recouvrant les endroits à corriger par du vernis correcteur blanc. Les illustrations; en dimension originale, préparées pour reproduction sont à insérer dans le texte. S'il s'avère nécessaire d'écrire de nouveau le manuscrit, soit complètement, soit en partie, la maison d'édition se déclare prête à verser à l'auteur, lors de la parution du volume, le montant des frais correspondants. Les auteurs recoivent 50 exemplaires gratuits.

Pour obtenir une reproduction optimale il est désirable que le texte dactylographié sur une page ne dépasse pas 26,5 cm en hauteur et 18 cm en largeur. Sur demande la maison d'edition met à la disposition des auteurs du papier spécialement préparé.

Les manuscrits en anglais, allemand ou français peuvent être adressés à Springer-Verlag, 6900 Heidelberg, Postfach 1780.